# THE ZEC GUIDE

## A Guide for Developing Zero Energy Communities

### By John Whitcomb

*AuthorHouse™*
*1663 Liberty Drive*
*Bloomington, IN 47403*
*www.authorhouse.com*
*Phone: 1-800-839-8640*

*TowardZero.org is a non-profit corporation*
*HTTP://WWW.TOWARDZERO.ORG*

*Published by AuthorHouse 11/25/2014*

*ISBN: 978-1-4969-5199-1 (sc)*
    *978-1-4969-5201-1 (e)*

*Library of Congress Control Number: 2014920033*

*Disclaimer: The ZEC Guide: A Guide for Developing Zero Energy Communities*
*offers guidance to developers, communities and businesses to assist them in developing*
*zero Energy Communities. The guidance and information in this guide is of a general*
*nature, and neither the author nor Goddard College is providing consulting services or*
*advice. You should not rely on this guide as an alternative to the services of qualified*
*professionals. Furthermore, no representations or warranties express, or implied,*
*are made by the author or Goddard College as to the contents of this guide.*

*Cover illustration by Jamey Stillings*

*This book is printed on acid-free paper.*

authorHOUSE®

# TABLE OF CONTENTS

I would like to share with you the course of my personal and professional life that brought me to this area of research. In June 2000, I attended the International Telecom Union Internet Symposium in Geneva, Switzerland. The purpose of this symposium was to bring together an international group of information technology and telecommunications leaders to focus on the impacts of the global Internet. This theme was of particular importance to me, because I was at the time, engaged to lead the development of a modern telecommunications and Internet control center for the Egypt Telecom Company.

Following the symposium and because of my engagement with the Egypt Telecom Company, the next leg of my trip took me to Cairo. There, I witnessed first-hand how the modernization of phone and Internet systems in Egypt could help its citizens improve their lifestyles and gain greater personal freedoms.

After the conference, and as I flew home over Libya, I reflected on an opposite outcome: how dictators and restrictive governments could exert control over the Internet, or worse—attack utilities and other important sources of power and information. I realized the fragility and vulnerability of data centers and control rooms that run power grids, telecom, transportation and financial networks. These realizations set the course for my next efforts to discover how best to reduce such risks in the United States.

As a result, for the next four years, I worked tirelessly to develop systems that would measure, manage, and overcome risks to data centers and control rooms. Of particular concern were those centers that operate critical aspects of government: defense, financial exchanges, telecommunications, electric power distribution, transportation, and Internet system utilities. These concerns led me to collaborations with government, military, and business sectors to define solutions that would make our nation more risk-averse. This initiative peaked in the aftermath of the 9/11 attack.

By 2004, I felt I had contributed to raising awareness, influencing policy, and creating new SEC industry regulations, providing tools and prototypes that changed industry practices. These formative experiences gave me faith and satisfaction in being a change agent.

In 2005, I became a consultant to Glendale Water & Power to develop a utility control room. This city-owned and operated municipal utility served a large community within the greater Los Angeles metropolitan region. Given my previous work and now living in the aftermath of 9/11, I was acutely aware of the issue of grid stability and intrigued by the emerging concept of "Smart Grids"—electric grids that use modern tools (like computers and cellular networks) to make them more reliable, interdependent, and efficient. Smart Grids also allow easier integration of energy efficiency measures or equipment, like renewable energy technologies and intelligent appliances that communicate with the grid and operate when electric demand is at its lowest.

I had already had a wide range of experience working with networks, as well as control centers, and had worked with governments and the private sector to improve power, transportation, telecommunication, and Internet networks. Developing Smart Grid systems seemed like an ideal application of my skills and a worthy challenge.

Following the Glendale project, I continued to research electrical utilities, meet with scientists, work with trade organizations, and attend symposiums. In 2008 I participated in a grant program for a green data center and in 2009 I joined the Rocky Mountain Smart Grid Consortium. Later that year, I applied for a position at the National Renewable Energy Laboratory (NREL). I received feedback that despite my career in this arena, I was not considered for the NREL position solely because I lacked a master's degree. Furthermore, I was told that the lack of a post-graduate degree would eliminate future positions of this type, as well.

As a result, obtaining a master's degree became important to me, and in 2010, I was accepted as a candidate for Goddard College's MA in the Sustainable Business and Communities Program, where I began to research how to upgrade the United States' outdated electric grid so that it could become the reliable Smart Grid of the 21st century. My subsequent research educated me regarding significant threats that the current electrical grid's energy sources pose to the environment. I studied more intensively and saw increased opportunities for energy conservation and the promise of renewable energy.

I concluded that utilities, the entrenched, regulated and often monopolistic organizations that provide electricity to most communities in the USA, are not adequately motivated to develop the necessary solutions to our increasing energy needs. Utilities, as traditionally conceived, profit from inefficiency and are motivated to increase their profits. Additionally, the prospect of overhauling their inadequate systems all at once seems to be financially risky and too overwhelming to current utilities.

I began to think about smaller-scale alternatives. I concluded that the most viable option for developing a more nationalized Smart Grid quickly was to form zero energy communities (ZECs), which reduce multiple threats from pollution, non-renewable energy, and the country's aging electrical grid. ZECs serve local energy needs, work within the larger grid, and serve as prototypes of a much larger, all-encompassing Smart Grid.

ZECs are a step beyond Zero Energy Buildings (ZEB), as scientists at the National Renewable Energy Laboratory explain:

> A net-zero energy community is one that has greatly reduced energy needs through efficiency gains such that the balance of energy for vehicles, thermal, and electrical energy within the community is met by renewable energy (Carlisle, AIA, VanGeet and Pless 2009)

As such, sustainable values are implicit in ZEC initiatives.

Benefits of ZECs include local economic development, environmental restoration, community building, and other economic and social benefits. ZECs engage community passion and inspire action from community members, businesses, institutions, developers, and planners. These factors converge to create what is often referred to as "triple bottom line" (TBL) benefits for people, planet and profit (*Slaper 2011*)

In my extensive study and research, I found considerable writing about the promise and practicality of ZECs, but I found little explaining how to develop ZEC projects. There have been no clear, comprehensive guides for developing a ZEC from the ground up. This guide was created to fill that gap. It provides a clear, concise framework and system for developing zero energy community projects.

# PREFACE

Zero energy communities (ZECs) are initiatives that allow groups to take a big step toward improving sustainability in their local environment and the world. *A Guide for Developing Zero Energy Communities* was designed to bring together developers, planners and community organizations, to inform them about the opportunities that ZECs offer, and to help them take action toward improving their own energy futures. ZEC formation can become a driving force that improves the economy, environment, society, organizations, institutions, and the world.

*A Guide for Developing Zero Energy Communities* provides comprehensive background and guidelines for the renovation of existing communities and the development of new communities toward zero energy community status.

This guide was created after decades of experience developing "built" environments that included infrastructure (including roads/bridges, water, sewer, power, gas, telecom), buildings (including homes, schools, hospitals, office, retail, mixed-use), and renewable energy systems (including solar, wind, geothermal, and biofuel). The guide provides encouragement regarding better methods of energy conservation by occupants (resource conservation, and recycling/up-cycling) and technical innovations aimed at realizing greater energy efficiency and cost reductions.

# ACKNOWLEDGMENTS

Many individuals contributed to this ambitious project:  the faculty at Goddard College, among them Ann Driscoll, Russ Gaskin, Giovanni Ciarlo, Richard Schramm, and Ralph Lutts, and writing advisors Arianne Townshend, William King, Chris Meehan, Kathleen Pray, Margo and Lana Whitcomb, and David Andrews, along with second-reader, Ellie Epp.

My good friends Steve Senk, Kenneth Witt, Robert Welch and Howard Schirmer each provided constant support and encouragement. Those who participated as interviewees contributed invaluable insight and resources to this research.

I also want to acknowledge the support of Design Workshop and Lowry Redevelopment Authority for their assistance in bringing ideas included in this guide to life at the Lowry ZEC development in Denver, which was being envisioned and built as this guide was developed.

William E. King, Chief Editor; Nancy Hutchins, Technical Editor, Renee Forsythe, Book Formatter and Jom Naknakorn, Graphic Artist were all instrumental in the development and publishing of this book.

Microsoft supported this project with Office and Office Smart Art software for graphics art, online pictures, and very supportive technicians from the US, India, the Philippines, and Central America.

My wonderful family provided me with an abundance of love, encouragement, assistance, and incredible patience as I researched and wrote. Every family member helped me in some way to make this book possible. Thank you all, especially my dear wife Lana.

# INTERVIEWS

**Alison Wise**, Principal, Wise Strategies, energy strategy renewable energy

**Arthur Hirsch**, Owner at Terra Logic, sustainability-engineering consultant

**Amory Lovins**, Chief Scientist and Founder at the Rocky Mountain Institute

**Bernays T. (Buz) Barclay**, Investment banker for power, renewable energy, infrastructure; legal counsel to entrepreneurs and project developers

**Jeff Lyng**, Senior Policy Advisor, Center for the New Energy Economy, Colorado State University

**David Andrews**, Project Manager, Lowry Development Authority

**David Driskell**, Executive Director, Community Planning and Sustainability, City of Boulder, Colorado

**Dennis Paoletti**, Acoustician, San Francisco Bay Area

**Fran Treplitz**, Energy Program Director, Green America

**Frank Ramirez**, Energy executive and entrepreneur

**John Keith**, President, Harvard Communities, green homebuilder

**John M. Prosser**, Professor Emeritus (ret.), School of Architecture / Urban Design, University of Colorado

**Joshua Pollock**, Goddard College graduate student and social media expert

**Kelly Crandall**, Sustainability Specialist, City of Boulder, Colorado

**Mark Dameron**, Chief Marketing Officer, EquityLock Solutions, Inc.

**Mike Ryan**, President of PanTera Energy, L.L.C, a geothermal utility

**Montgomery Force**, Owner of Force Consulting and Executive Director at Lowry Redevelopment Authority

**Peter Asmus**, Writer and Analyst at Pike Research and Navigant Consulting

**Paul Thompson,** Patent Agent, Cochran, Freund and Young L.L.C.

**Robert Welch**, Chief Technology Officer, TowardZero.org

**Sarah Bobrow Williams**, Community Development Expert, Goddard College

**Skip Spensley**, Community Development Consultant, Prof. University of Denver

**Sunil Cherian**, Chief Executive Officer, Spirea, Smart Grid entrepreneur

**Timothy Collins, Sr.,** CEO, KleenSpeed Technologies

**Todd Johnson**, Partner, Design Workshop, Inc., land planning and design

**Victor Olgyay,** Principal at the Rocky Mountain Institute

# INTRODUCTION

Would you sleep better at night knowing your children's alarm clocks will always have power—and not just power—but clean, reliable energy? You are not alone. Increasing numbers of people favor energy efficient, environmentally restorative, and economically sound approaches to energy. They also want to play a part in sustainable local economic development. Still others aspire to help accelerate the advancement and deployment of a non-polluting national Smart Grid. All of these goals can be realized through development of zero energy community projects, enhancing the lives of individuals, communities, and businesses. A leader in the conversation about the nation's energy strategy, former Colorado Governor Bill Ritter, who is the Director for the Center for the New Energy Economy, Colorado State University stated:

> *The American people want renewable energy, even if they have to pay more for it.*
> *(Governor Bill Ritter 2012)*

One of the quickest and easiest ways to move the US to adopt renewable energy and simultaneously embrace energy efficiency is through development of zero energy communities. ZECs use as much energy as their renewable energy resources produce. A growing body of information indicates an emerging trend toward developing ZECs in the United States.

> *A net-zero energy community (ZEC) is one that has greatly reduced energy*
> *needs through efficiency gains such that the balance of energy for vehicles,*
> *thermal, and electrical energy within the community is met by renewable energy.*
> *(Carlisle, AIA, VanGeet and Pless 2009)*

ZECs largely meet onsite energy demand with local energy supplies that are efficient, reliable, non-polluting, and affordable to all. Not only do ZECs provide energy independence, they conserve energy and reduce or eliminate fossil fuel emissions at the community level. The growth of ZECs is among the best and fastest ways to move the US toward making its national electric supply more reliable, environmentally improved, and economical.

The widespread development of ZECs will foster energy conservation and the transition to clean, locally produced, renewable energy with which to power homes, buildings and charge electric vehicles. Cost benefits of ZECs are realizable through up-front financing that spreads the cost of improving building energy efficiency and purchasing renewable energy generating equipment, such as solar panels and wind generators, over many years.

> *The result of this type of up-front expenditure yields long-term savings by*
> *creating consistent returns for financiers and predictable energy costs for the*
> *community (Lovins, Reinventing Fire: Bold Business Solutions for the New Energy*
> *Era 2011) The primary financial benefits of a ZEC occur because of utilizing*
> *renewable energy sources (that displace all fuel costs) and long-term avoidance of*
> *steadily increasing costs of fossil fuel (Yergin 2011)*

The detailed process in this guide is intended for use by any institution, business, nonprofit, real estate developer or citizen's group that seeks to plan, and complete, the development of a ZEC. Depending upon the ZEC project size and complexity, ZEC planners may need to engage and manage the efforts of additional specialists in architecture, design, planning, project management, energy, law, or other areas of expertise to aid in project development and/or meet state and local requirements. Those interested in energy security and/or reducing use of fossil fuels may also use this research-based guide and the many resources provided herein.

This guide was designed in accordance with the definition of ZECs developed by the National Renewable Energy Laboratory (NREL) to allow readers to take on the development of a ZEC with the hope that groups throughout the US will reduce America's dependence on foreign energy, improve health and the environment, and diversify the base of national energy security–one ZEC project at a time.

This guide will answer questions like: What is a ZEC? What are its goals? How do ZECs reduce energy consumption and increase local energy production and economic growth? How are ZECs financed? What are the major challenges to designing and executing these initiatives? Where has this worked? How do you ensure success?

The appendix includes numerous case studies of Zero Energy Buildings and community energy projects. While there is no comprehensive list of current ZEC projects, supplementary research is being undertaken by the author to create and manage such a list.

> *Sustainability is an inevitable outcome. Conservation is essential. Developers need more education. We all need to understand the resistance. The situation will never return to the status quo. People want a meaningful place to live, just like kids today want a meaningful job. (Johnson 2013)*

# ABOUT THE ZEC CONCEPT

Zero energy communities meet onsite energy demand with local energy supplies that are efficient, reliable, non-polluting and affordable. Not only do ZECs provide energy independence, they also conserve energy and reduce or eliminate fossil fuel emissions at the community level.

A ZEC community is designed to produce as much energy as it uses. It is developed by adhering to a program and plan that addresses the buildings and infrastructure (roads, bridges water, sewers, etc.) that are referred to as the "built environment." While a ZEC does not encompass a sustainable transportation program (see Section 10: Electric Vehicles), NREL requires the accommodation of electric vehicle charging stations at all parking places within a ZEC.

$$\text{Energy Demand Forecast} - \left( \text{Onsite Energy Generation} + \text{Offsite Renewable Energy} + \text{Renewable Energy Credits} \right) = 0$$

*Figure 1 - FORMULA FOR A ZEC (Used by permission of Design Workshop, Inc. and Lowry Redevelopment Authority)*

The sources and uses of renewable energy add up to the zero goal. Achieving a *RESULT GREATER* than zero indicates the need to continue to optimize the energy balance. ZECs that make a best effort to reach zero, even if above zero, may still be certified as ZECs. Electric vehicles account for approximately fifteen percent of a community's energy use.

The formula for a ZEC may be used to evaluate the net-zero balance of community electricity (Watt-hours) and British Thermal Units (BTU). By running the two sets of numbers, ZEC planners gain greater awareness of how potential solutions can most effectively achieve the desired net-zero energy balance.

> *Zero energy communities are built upon principles and best practices, which include determining appropriate clean energy technologies and the consideration of local ecosystems. Zero energy communities are attainable in almost any circumstance when participating groups make an authentic, informed and sustained attempt toward a net-zero energy balance. Like other sustainability initiatives (paraphrasing NREL Senior Scientist Terry Penny) ZEC definitions are intended to encourage best efforts by developers. (Penny 2012)*

The Department of Energy has achieved many successful national energy improvements, especially in the past thirty years. Those programs provided leadership to the business community, which has actively developed and improved technology, cost viability and financing models for energy efficiency, renewable energy and Smart Grid technologies.

*In the past five years, partly due to the energy department's support—through ARRA (The Stimulus) and other funding — America has deployed the largest fleet of microgrids in the world (Navigant Consulting 2013). Microgrids utilize renewable energy, energy storage and "Smart Grid" control and automation systems. This is relevant because Zero energy communities (ZECs) are microgrid applications and are, when on-grid, also part of Smart Grid. (Navigant Consulting 2013)*

*Global climate change is a driver of the quest for a Smart Grid that utilizes renewable energy. Whatever varied beliefs about climate change exist, the United States government is currently funding Smart Grid research and has suggested that a carbon tax will be good for the economy and environment, and can be accomplished without undue impact to low-income Americans. (Lester 2013)*

*At the same time, power costs have increased, and will continue to rise according to the US Energy Information Administration (EIA) and Department of Energy (DOE), which both project increases in the cost of retail electricity to rise between 38% and 101% between 2015 and 2050. (Sunset Vision Study, US Department of Energy 2012)*

The ZEC approach, however, avoids rising energy costs and price volatility by relying on renewable energy resources. With significant advances in renewable energy and energy efficiency improvements occurring since the year 2000, the costs of implementing and operating the technologies are reduced, while equipment life-cycles are longer, making the case for ZECs even more attractive.

### What the HECK is a ZEC?

A ZEC is a land site that achieves a net-energy balance within its defined geographic border. In a ZEC, energy is conserved, renewable energy supplies are deployed, and electric-vehicle charging is accommodated. The result is a community (or campus) that relies on renewable energy to meet its average annual energy demand.

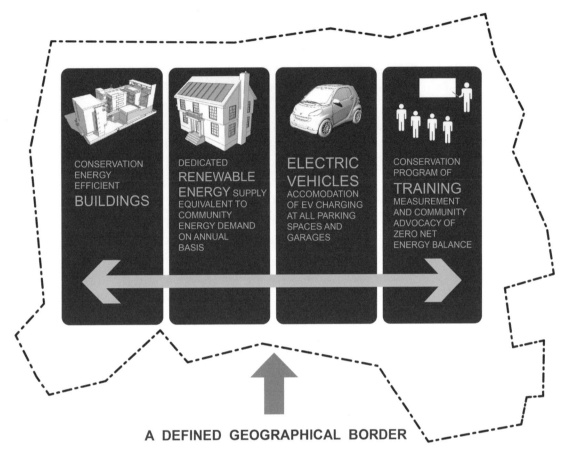

*Figure 2 - DIAGRAM OF A ZERO ENERGY COMMUNITY THAT PROVIDES ENERGY CONSERVATION, RENEWABLE ENERGY SUPPLY, ELECTRIC VEHICLE PLUG-INS AND LEADERSHIP WITHIN A DEFINED GEOGRAPHICAL BORDER*

Today, ZECs have been limited in scope to neighborhoods, block grids of cities, campuses and military bases. In the future, large ZECs could span broad regions and encompass large cities.

A ZEC can be implemented by developing a community energy policy that has a foundation of energy conservation and addresses the net-zero balance of energy supply and demand. Because electric-vehicles (EVs) will change the energy requirements of ZECs in the foreseeable future, the electricity required to charge EVs is contemplated in the net-zero equation. A ZEC may interconnect with the power grid and rely on that grid to support instantaneous imbalances in supply/demand, provided the annual net-zero energy balance can be achieved.

### Why Use NREL's Definition of a Zero Net Energy Community?

The National Renewable Energy Laboratory (NREL) is a federally funded US laboratory of research science and deployment and commercialization focused on renewable energy. Although definitions for Zero Energy Buildings (ZEB) and other green buildings already exist and are commonly accepted, the ZEC is a new concept, and the first definition of a ZEC was published in 2009 by the National Renewable Energy Laboratory.

This guide is among the first—if not *the* first—published ZEC guide, and is based on the tenets of the NREL's definition of a ZEC. The NREL definition of a ZEC is thoughtful, practical, technically correct and timely. However, a definition is not a guide, and it does not deal with the detailed processes for designing, implementing, and living in a ZEC; therefore, this guide was developed as a tool to help inform project communities, developers and planners in achieving ZEC status.

Under NREL's definition of a ZEC (see below), it must incorporate measurable milestones that assess value, judge authenticity, and allow comparison of ZEC projects. NREL's ZEC definition is an authoritative introduction to ZECs for a number of reasons, among them:

***NREL is a global leader in renewable energy and energy conservation research.***

NREL has created, tested and evaluated numerous renewable energy technologies.

From its inception in the late 1970s, NREL has supported public-private research and development, deployment and commercialization of renewable energy and energy efficiency solutions.

***NREL is reliably expert in the areas of science affecting ZECs.***

The ZEC framework is an attractive solution to addressing the sustainability at a meaningful scale because it addresses several aspects of the energy dilemma simultaneously:

ZECs support the adoption of plug-in electric vehicles (EVs), thereby providing a means to shift energy used in transportation, the largest use of energy, from fossil fuel to the ZEC's renewable energy.

ZECs shift energy used to power buildings, the second largest use of energy, from fossil fuel to the ZEC's renewable energy, and greatly conserve that energy utilization.

ZECs provide a means to engage communities in the conservation of energy by limiting their consumption.

A ZEC must achieve a net-zero energy balance on an annualized basis, meaning it must produce or purchase as much renewable energy as it uses during each year. The balance is attained through implementing energy efficiency or including locally produced renewable energy, purchased renewable energy, and/or purchased renewable energy credits (RECs).

Since the supply of renewable energy may be irregular, and peoples' energy usage also varies on an hourly, daily, weekly and monthly basis, there are likely to be imbalances in supply and demand. An on-grid ZEC depends on power supplied to a ZEC by the electric grid. That power will smooth those fluctuations in over-supply and under-supply. An off-grid ZEC will support its own peak energy load on an instantaneous basis. For this reason, the off-grid ZEC must have more renewable energy and energy storage, and therefore, an off-grid type ZEC is considerably more expensive.

NREL's ZEC definition is realistic and considers financial costs within the normal project hurdle rate. Under the definition, communities are not expected to spend exorbitant amounts to comply with onsite renewable energy, for instance. It allows that technical performance is limited by what is technologically available and reasonably affordable. Financial investments in a ZEC program are not expected to exceed typical hurdle rates of the development industry and renewable energy developers.

NREL also recognizes that electric cars and plug-in hybrid cars will soon influence community energy system designs and need to be incorporated into ZEC planning.

### Using NREL's Definitions and Criteria

*NREL defines a ZEC thus:  A net Zero Energy Community (ZEC) is one that has greatly reduced energy needs through efficiency gains such that the balance of energy for vehicles, thermal, and electrical energy within the community is met by renewable energy. (Carlisle, AIA, VanGeet and Pless 2009)*

### NREL stipulates that a ZEC must meet three criteria:

- First, there must be a defined boundary for the ZEC. This may be the city limits, the neighborhood property line, the borders of an educational, health or corporate campus—or the shoreline of an island.
- Second, the ZEC planners are required to establish measurable milestones that the community can plan toward, and aspire to, over a period of many years, narrowing the gap on net-zero over time. The definition is meant to encourage developers and campus and community planners to set out and drive the transition.
- Finally, the ZEC plan incorporates charging for plug-in electric vehicles (EV) at all the parking spaces within the boundary.

### Using NREL's Classifications of ZEC's

According to NREL, there are several classifications for ZECs. These are designed to rank the effectiveness of the ZEC in reducing energy consumption, increasing use of renewable energy, and restoring the environment. These classifications are listed in order of their desirability:

- It is most desirable if the ZEC incorporates renewable energy from otherwise unusable Brownfield sites.
- The next most desirable outcome is that a ZEC offset all of its annual energy use from renewable energy generated within the community's boundaries.
- A ZEC that offsets all of its annual energy use from renewable energy generated outside the community's boundaries is still a worthy goal.

A ZEC that offsets all of its annual energy use primarily through the purchase of new, off-site Renewable Energy Certificates (RECs) is placed at the lowest end of the ZEC classification, but is still considered a good achievement.

### Using Electric Vehicles Charged with Renewable Energy

The NREL definition of a ZEC requires charging stations at parking places to accommodate electric vehicles. These vehicles, when charged from the ZEC's renewable energy system, become carbon neutral. The ZEC accommodates EVs, but that does not mean that all of those in the ZEC community own only EVs. Hopefully, people will all upgrade to EVs when they have to replace their current cars and trucks in the future.

*According to NREL, a ZEC is created as follows: The hierarchical approach emphasizes using all possible cost-effective, energy efficiency and demand-avoidance strategies first and then using renewable energy sources and technologies that are located in three places. (Brownfield, Greenfield and remote sites - see Figure 3) (Carlisle, AIA, VanGeet and Pless 2009)*

NATIONAL RENEWABLE ENERGY LABORATORY

| National Renewable Energy Laboratory (NREL) Community Energy Efficiency And Renewable Energy Supply | |
| --- | --- |
| Option Number | Option Name |
| 0 | Energy Efficiency and Energy Demand Reduction |
| 1 | Use Renewable Energy in the Built Environment & on unusable Brownfield Sites |
| 2a | Use Renewable Energy on Community Greenfield Sites (A Greenfield site is a site that has not been previously developed or built on, and which could support open space, habitat or agriculture) |
| 2b | Use Renewable Energy Generated Off-site, On-site |
| 3 | Purchase New Off-site Renewable Energy Credits (REC's) |

*Figure 3 - NREL'S GRADING FOR THE VALUE OF ZERO ENERGY COMMUNITY TYPES*

### Conclusion

The NREL definition of a ZEC provides a framework that makes it practical to develop a ZEC in any situation. Given the adoption of innovations like Virtual Net Metering, ZECs may someday be developed using virtual borders formed by ownership or circles of people connected by social media. There is room for all of us to imagine and better re-shape the ZEC of tomorrow.

# SECTION 2
# LIVING AND WORKING IN A ZEC

There will be economic, practical, health-promoting, aesthetic and social benefits for those who live or work within a ZEC. These community members can be proud participants in projects that will help to redefine how the world can use clean energy for transportation, buildings and industrial processes. ZECs are designed to solve energy problems, while also providing a high quality of life for their residents and workers that yields benefits beyond energy and creates other value.

### Health, Productivity and Well-being

The Center for Disease Control (CDC) has determined that living in a sustainable home and community provides improved health over living in conventional communities (Centers for Disease Control and Prevention 2010). That may mean that medical and other health-related costs of community members in a ZEC would be lower when compared to those of peers not in a ZEC.

*The National Health Institute (NIH) performed studies of worker productivity, absenteeism and feelings of well-being before and after those workers were relocated to workplaces in green buildings, concluding that improved indoor air quality (IAQ) contributed to reductions in perceived absenteeism and work hours affected by asthma, respiratory allergies, depression and stress, and to self-reported improvements in productivity. (Centers for Disease Control 2013) These facts are validated by other studies: preliminary findings indicate that green buildings may positively affect public health. (Singh, et al. 2010)*

Reliable Energy: Americans have become accustomed to a limitless supply of inexpensive electricity and fuel, but this trend is turning, as the impacts of widespread blackouts and the price of energy steadily increase. A ZEC stabilizes electrical systems and greatly reduces risk of blackout or brownout, and this reliability will be beneficial when there is a widespread or sustained power outage.

Community Benefits: ZECs can provide a community with independence and a level of control regarding the cost and environmental quality of their energy. A ZEC reduces energy consumption dramatically, and provides primary power using renewable energy, which, at least in the long-term, will cost less to produce than any other known alternative.

- ZECs provide green buildings, alternative transportation and renewable energy, which measurably reduce environmental impacts and risks.
- Because of the environmental quality of the ZEC, occupants should enjoy better health and reduced costs of healthcare.
- All of those who work or live in a ZEC will enjoy the cost-savings potential of electric vehicles, which will generally be less expensive to operate when they are charged-up using the ZEC's renewable energy.
- Worker productivity can be improved in a ZEC because its green buildings cause a material reduction of absenteeism, depression, and stress that provide improvements in productivity, and create financial and other value for employers who operate offices, factories, stores and provide services in a ZEC.

- Commercial users of multi-tenant buildings share the maintenance expenses related to their common-areas that include energy cost. The reduced life-cycle operating costs for that common property energy can also benefit from energy-efficient lighting and irrigation systems that reduce the building's operating expenses. In the same way, a community's reduction of energy used in public spaces and buildings may be a benefit to the operating cost that can result in lower taxes and homeowner association fees, as well as competition for school funds.

- *Buyers of real estate located in a ZEC can realize mortgage qualification amounts that are higher, and interest rates that are lower, due to lenders' consideration of lower utility bills, and therefore, an increased ability for borrowers to pay their mortgages. (Dameron 2013)*

- Property owners in a ZEC, whether they are sellers or landlords, should experience improved marketability, higher appraisal prices, higher market rents, and yield some beneficial effects from the financing benefits provided to the buyers. There is an expectation that the homes, lots and buildings located in a ZEC will retain value (or appreciate) at a better rate than their equivalents elsewhere.

- Value accrues to residential and commercial property. Owners and renters of ZEC property can benefit from lower utility bills for the offices or apartments they own or rent. This saving may reduce the consumption and cost of electricity, natural gas, oil and other fuels.

- Drivers may enjoy greater availability of PEV charging stations.

- Parking spaces (usually located beneath solar canopies) provide shelter from sun, rain and snow in ZEC parking lots.

- All users of ZEC property benefit from the cost reduction and value increases consistent with ZEC property.

- Since ZECs are designed for energy efficiency, homeowners, renters and businesses within the ZEC should realize lower energy expenses.

- Residents, businesses and institutions located in a ZEC reduce water consumption because energy-efficient appliances require less water.

- In addition to utility cost savings, ZEC occupants may also receive purchase-cost rebates, tax credits and other incentives offered by the ZEC, the federal or local government, or local utilities.

- Sellers of ZEC property are likely to realize an increased property resale value compared to equivalent property not located in a ZEC because buyers of ZEC properties can verify a history of lower utility costs, reducing the buyer's overall cost of ownership for the property.

*Figure 4 - ZECS IMPROVE THEIR LOCAL ECONOMY, THE ENVIRONMENT, AND LIVING CONDITIONS.  (Photograph courtesy of Jamey Stillings)*

### Other Benefits of a ZEC

Getting around a ZEC: Walkable access to markets, available car sharing, and accessible public transit can all result in cost savings for the occupants and users of the ZEC.

*High efficiency lighting in a ZEC: Lighting is better and less expensive to operate. Since high-efficiency LED light has an inherently longer life-cycle than other lights, LEDs used for office, household and street lighting will greatly reduce energy, saving money over other lighting forms. LED streetlights are often solar powered and can feasibly be installed without wiring them to the grid—reducing infrastructure and deployment costs. In addition, studies have shown that LED-based lamps offer better light clarity at night. (Shih-Fang Liao, et. al. 2009)*

This higher quality of light means that the artificial light from LEDs will stimulate the rods within the human eye in dark environments, compared with illumination from HID, mercury vapor or fluorescent light sources. The result is better night vision with lower ambient levels of outdoor lighting.

Acoustic benefits of a ZEC: Controlling environmental noise is not often considered in the hectic progress-driven world we live in. Since the 1950s, air conditioning of commercial office buildings and in-wall air conditioning units in residential buildings have plagued the environment, contributed significantly to the excessive use of energy and generation of carbon, and have created noise that has robbed us all of the peace and tranquility we desire—all in an attempt to bring comfort to the modern world.

However, a reduction in environmental noise is another benefit of living in a ZEC. Solar and/or energy efficient heating, ventilation, and air conditioning (HVAC) systems lower interior and exterior operating noise levels. Well-insulated commercial and residential (multifamily as well as single family) buildings offer reduced interior and exterior noise when compared to conventional units typically using noisy equipment on the building exterior, and requiring more forced air on interior zones. This is particularly true when solar energy, geothermal heat pumps, and other specially designed sustainable products are utilized.

> *Quieter offices, stores, homes, and balconies in a community will improve productivity (in the workplace), provide peace of mind, reduce stress and improve our quality of life. These are subtle but important benefits of living in a ZEC. (Paoletti 2013)*

Air quality in a ZEC: Air quality in ZEC buildings may be improved by using ecologically sound materials and efficient construction methods that minimize "off-gassing" which refers to the tendency of the materials to release toxins into the building air.

In a walkable or bikeable community, inhabitants are encouraged to walk or ride to local markets, use public transit and drive PEVs. All of these behaviors reduce emissions into the community's indoor and outdoor air, leading to additional health benefits.

Incorporating high percentages of green space and green roof structures typical of sustainable communities further enhances local air quality. The pleasant outside environment may also encourage residents to enjoy more outdoor living and fresh air, again encouraging a more active, healthier lifestyle.

> *Transportation efficiency in a ZEC: Sustainable transportation systems are designed to reduce community energy demands while simultaneously harnessing, and making more controllable, the means of generating that same energy. The term "eco-cities" describes a sustainable transportation system: We need not only to develop foot, bicycle, and public transportation; we also need to put transportation into the land use context. Emphasize providing access through ecological urban and architectural design and planning of city layout. (Register 2012)*

Imagine a world with convenient and more capable transportation alternatives that would not pollute the environment, would be more convenient, and would be more capable in terms of comfort, range and speed, than current transportation options. Many cities are in the process of computerizing traffic lights in order to make more efficient movement patterns for traffic, thereby reducing pollution, wasted time and energy.

Imagine returning home from the office in a bus, car or van-pool vehicle that accelerates like a sports car, makes no engine noise, has no tailpipe exhaust and requires no foreign energy to operate. Moreover, the electricity used to power these vehicles has come from fossil-free solar, wind and other renewable sources. The cost of operating these vehicles is drastically reduced because renewable fuel is free.

Choose a ZEC for sustainability: I define sustainability as a driving force of change that improves the economy, environment, and social condition of communities, organizations, institutions and the world. ZEC initiatives embody sustainability.

> *Benefits of ZECs include local economic development, environmental restoration, community building and other social benefits. These benefits align to create what is often referred to as the triple bottom line (TBL): people, planet, profit. (The Economist 2009)*

ZECs capture passion and inspire action from community members, businesses, institutions, developers and planners.

Intangible benefits: People who help to create ZECs, and who live and work in ZECs, all play a pioneering role in America's new energy economy. The admirable ability to be a part of a community that understands opportunity and puts energy solutions into action will be rewarded in countless ways. The ability to solve these problems while bestowing economic benefit on the participants might make ZEC residents into a new version of American heroes.

Sense of community: Creating a ZEC brings together communities, planners and involved organizations to learn about and take action regarding their own energy future.

> *Zero energy communities are initiatives that allow groups to take a big step toward improving sustainability in their local environment, as well as the world overall. The positive impacts of local initiatives, like ZECs, and the manifold impacts localism makes to local economies, are examined in* Ecology of Place: Planning for Environment, Economy, and Community. *(Beatley and Manning 1997)*

Sense of purpose and accomplishment: Whether the ZEC is a corporate campus, a mixed-use community, or a school, there is a brand value that accrues to all who participate in it. Moreover, for those people, the connections built with other forward-thinking people, and new ideas, should increase benefits exponentially. ZECs serve as a catalyst for a process that brings positive changes, both economical and environmental, as well as in terms of social citizenry.

Social equity: Providing an opportunity for social diversity within a ZEC is not a requirement. As with the creation of any intentional community, ZECs present an opportunity for inclusion and diversity. By integrating people of different ethnicities and socio-economic strata, a ZEC can potentially create a multicultural village.

> *For example, your ZEC might have a cultural district oriented toward a particular ethnicity by providing a place for an international school, incorporating a retirement home or low-income housing, and including a church, mosque, synagogue and/or temple or other multicultural feature. Dr. Martin Luther King reminds us that "All of us are caught in an inescapable network of mutuality, tied in a single garment of destiny. Whatever affects one directly affects all indirectly." (Goodman 2011)*

### Conclusion

All combined, the features of a ZEC make for an incredible living and working environment. Whether your interest is in a location for your business or institution, or a place to reside, a ZEC's benefits will be valuable. Financial investments in ZECs are more likely to return gains, as the costs of energy and healthcare continue to rise. A ZEC creates value for the community, and those who live, shop or work there do so in healthier green buildings.

ZEC participants can feel pride in being part of a community that works together to conserve energy and use renewable energy. They can experience pride of ownership and the comfort of knowing they are helping to create a more secure, sustainable and diversified community. Individuals can enjoy the freedom to enjoy life more, both indoors and out. They can take a fresh breath, expect a quiet in the air, and walk a path or ride a bike, as they enjoy an aesthetic that blends nature and the "built environment" together without infringing on the environment. Each person can settle into a place full of people who want to live better and enjoy peace of mind.

The ZEC paves the way to a new energy economy in America by supporting the conversion of gasoline-powered cars to electric vehicles, energy-consuming buildings to zero-energy buildings, and operating these with the endless and economical supply of renewable energy.

# SECTION 3
# FINANCIAL CONSIDERATIONS

"Zeconomics" refers to alternative practices and principles of marketing and financing in support of Zero Energy Community (ZEC) development. If there is to be a new paradigm of energy systems in a community, there will be significant costs and returns to be understood. Former US Vice-President Al Gore said:

> The more oil and coal we use, the more expensive it gets. The more solar energy and wind energy we use, the cheaper it gets because it benefits from automation. In 2010, world investments in renewable energy exceeded investment in fossil fuel. (Gore 2013)

The investment a community makes in upgrading the energy efficiency of buildings and deploying renewable energy within a ZEC project may be significant. Conversely, the energy cost savings eventually return the initial investment. The time required to repay the investment in a ZEC upgrade through savings realized as a result of that investment is called the "breakeven period." Following the breakeven period, the investment will yield a continuum of cost savings.

An additional consideration of the investment is that of property value. The market value of the real estate within a ZEC may change based upon the real or perceived value of green buildings and a ZEC designation. The initial cost to create green buildings and infrastructure in a ZEC community is greater than the cost of traditional building practices that rely on traditional gas and electric utilities. The cost delta is referred to as a "green premium" and there are many considerations to make in order to justify the increased investment:

### Who Takes the Financial Risk Associated with a ZEC?

ZECs are a fairly new phenomena, and they have not yet established a record of producing good financial returns for the developers, homeowners and businesses that make investments in ZEC property development. There is solid evidence that green buildings create financial value, but the idea that a community full of green buildings will succeed in creating strong financial returns is only a projection today.

> Homeowners who invest in purchasing a home in a ZEC, or a neighborhood in the process of becoming a ZEC, face uncertainty. Are these homes worth more than the conventional houses of today? What about tomorrow? When will the ZEC value be apparent? John Keith, President of Harvard Communities, Denver's leading green homebuilder, comments: homebuilders make large investments in land and speculate on the construction of new homes. We must invest cautiously in green features that will produce a financial return because there is only a small profit margin, and half the customers do not care about efficient solar equipped houses. (Keith 2013)

Likewise, existing homeowners who invest in improving their home energy efficiency and renewable energy systems as part of a ZEC initiative through renovation, also face uncertainty. When will the financial benefits occur?

Homebuilders building model homes and speculative homes in a ZEC neighborhood are making a significant bet that a marginal increase in building cost will result in an increased sales price and similar sales cycle duration. Additionally, the homebuilder is employing new technologies and methodologies in the construction technique, adding additional technical, execution and warranty risk.

Real estate developers desiring to develop a new Greenfield development of a ZEC must create building lots that favor solar rooftops. This means that building lots must be oriented so that the buildings face the path of the sun, which means that in the US, all buildings face the south. This results in suboptimal layouts relative to the number of lots that can possibly be sold from a given plat without constraint. In addition to the reduced yield of land use, the economic efficiency of street, bridge, water, sewer, lighting and communications infrastructure are all areas of greater costs and investment risk. The premium value of solar oriented building lots must produce higher prices in order to make up for having fewer developable lots. Montgomery Force, Owner of Force Consulting and Executive Director at Lowry Redevelopment Authority, explains why a developer contemplating a ZEC must ask if the customers in the market will pay:

> *There are technology risks associated with renewable energy that we cannot assume we understand today. In the past when we built fiber-optic communications directly to each house in the neighborhood, we found that fiber-optic was not the market requirement any longer. The technologies for high-speed data could then be delivered over copper cables - our investment in fiber-optic technology did not pay off. Will investment in renewable energy and energy efficiency pay off? This is a dilemma for developers, one that we are still learning about. (Force 2013)*

### The Role of Utility Companies

Are utility companies the logical investors and operators of ZECs? They have the expertise required, the capital to invest and the customers to serve. Unfortunately, a utility company sees net-zero energy as net-zero revenue. The business model of utility companies would have to change materially for them to succeed in the ZEC business.

> *American utility companies use sophisticated financial mechanisms to invest even more capital to meet growing demands for the reliability of electricity. Many utility companies are monopolies who are rewarded for investing capital and raising the price of electricity. The electrical power industry is increasingly producing a new breed of companies like Liberty Power in Florida, that have sold over a billion dollars' worth of green electricity to US consumers and businesses. Other utility companies are succeeding in shifting their focus to the needs of an educated population. More companies will enter this space. Capital is available for large projects. There is very smart money making investment in the sector. (Ramirez 2013)*

### How Is Risk Ameliorated?

Risk is relative. The confidence of an investor is proportional to that investor's experience in gauging those risks. Expert investors have skill at determining probable outcomes, vetting the risks, and understanding the financial consequences. Investors operate in specific areas of the market where their expertise in vetting the risk and potential upside of investment is competitive.

ZEC risks are partly taken by the real estate developers, partly by the consumers, partly by the homebuilders, and possibly the utility companies and renewable energy producers, and definitely by the financiers. Each of these groups needs to be educated regarding the cost and benefits of a ZEC. By considering the risk tolerance of each group, the ZEC planning team can appropriately divide the risks among the parties who are most able to manage them.

*Understanding the Cost of ZEC Improvements*

The ZEC planning team can determine the magnitude of cost for achieving energy conservation and renewable energy, the key questions being:

(1.) How can the ZEC encourage homeowners to invest in energy efficiency renovations for their existing homes? The ZEC team can enter into dialogues with architects, builders, and energy performance contractors (EPC) to understand the typical costs given the efficiency goal (e. g., HERS scores of 40-50) and the age and construction typical to homes in the community. This Zeconomic factor is unique to each community because construction-related costs vary, as do regional energy costs. Home construction types also vary and the local climate has a significant impact. By determining an average cost per square foot for weatherization and estimating the financial return expected from energy savings and available financing alternatives, ZEC planners can help homeowners understand and balance cost/risk ratios.

(2.) How can the ZEC planners for a Greenfield ZEC convince the builders to invest in building houses and buildings that are highly energy efficient? When the planners of the Lowry ZEC asked seventeen leading homebuilders about this perceived extra cost, they unanimously responded by affirming that the additional costs required to create energy efficient homes would be minimal, a new market requirement, and that their firms would therefore offer homes with ratings between HERS 40-49 at Lowry. Many of the national homebuilders had already made energy efficiency a standard in their products. The investment is justified to these homebuilders, who take the investment risk by building green buildings that sell for an amount that is only slightly higher than a typical home.

(3.) How does the real estate developer become comfortable with the ZEC? The real estate developer invests in creating energy design standards for a ZEC and related infrastructure development costs in hopes that the lots will be sold at a higher price. Developers are put at ease when they understand they do not have to finance renewable energy equipment, when homebuilders agree to develop and market the ZEC, and when commercial developers agree to build any required retail, apartments, condominiums and office spaces with energy efficiency features (e.g., Energy Star III).

(4.) Who invests in required solar panels? The cost of solar equipment is material (often tens of thousands of dollars per home) and perhaps the biggest financial consideration of the ZEC. There are an expanding number of options available:

A. The homeowner can purchase solar panels using financing or leasing plans. This way the homeowner uses the immediate savings in utility costs to pay for the panels, and once they are paid off, the consumer enjoys free electricity for life.

B. The homebuilder provides options for solar panels to the homebuyer and the cost is then included in the mortgage financing for the home.

C. A solar developer, an independent power producer (IPP) or energy service company (ESCO) will want to design, install and operate those systems under a power-purchase agreement (PPA) in the case of large solar gardens and solar installed upon commercial buildings, and other large equipment such as generators.

In all cases, there are considerations to be made regarding the prevailing interest rates, availability of capital, and the anticipated escalation costs for conventional energy. Additionally, there may be incentives available in the form of low-interest loans, tax-credits, rebates and sales of renewable energy credits (RECs).

Also nested within the Zeconomics equation are the issues of timing. When will homeowners pull the trigger on renovation or installation? When will homes and buildings actually be built and occupied with energy users? What is the phasing plan for large ZEC development? How does the investor or financier know when their investment will be repaid?

Other considerations with a community operating model include determining who the energy service company (ESCO) would be contracting with and deciding the ownership of the energy infrastructure. What would happen if the ESCO or the community defaulted? What is the credit worthiness of each entity? In the case of a neighborhood, how would the energy bill be divided, and what would be the recourse of the ESCO if certain homeowners failed to pay their energy bill?

Real estate developers are familiar with a similar dilemma related to the maintenance of "common areas" within a development, such as entrances, paths, parks and recreation areas. A Homeowner's Association (HOA) is usually formed at the outset of a new neighborhood, which through its dues, collects the funds necessary to provide maintenance for the development. HOAs have other authorities as well, and they take a security interest in every property to avoid risk.

For the Lowry ZEC in Denver, the Redevelopment Authority understood that there would need to be an entity chartered to represent the energy interest of the ZEC. They envisioned a Sustainable Homeowners Association (SHOA) to serve the community needs in regard to a broad set of sustainability goals that include the energy matters of the ZEC. Through a combination of dues, and payments for energy and other services, the SHOA collects funds to provide maintenance and community sustainability services. The SHOA serves as the intermediary between the community and entities such as ESCOs, because like an HOA, the SHOA will have other authorities and holds a security interest in every property in the community.

Another aspect of Zeconomics is related to a shared ownership of renewable energy equipment systems. A ZEC may invest in a centralized system of renewable energy that is distributed to the ZEC occupants. When this occurs, the payments from many residents are aggregated to pay for the energy or equipment financing. This may be based on the actual sub-meter readings for all occupants, or other formulaic calculations, such as a pro-rata basis. In regions where virtual net metering is legal, a ZEC may invest in, or purchase energy from, a large renewable energy supply and that cost needs to be distributed to the occupants of the ZEC.

### Financing the Cost of ZEC Improvements

In a typical community (non-ZEC), it is the electrical utility company that invests in all necessary electrical equipment and provides all operational support and maintenance. The utility then bills the users of the energy according to meter readings. It would be great if utilities could be reconceived to offer ZEC services on the same basis, and I am hopeful this becomes the case in the future. After all, utility companies are sophisticated financial experts who understand energy transactions and can determine risk and reward, as well as operate to provide wealth to their shareholders.

Today, there are many ways to finance net-zero energy technologies. Many projects are made possible by having a community, business or institutional customer engage engineering and building firms to implement a solution and seek third-party financing. Those customers have choices in selecting their supplier of energy.

Some ZECs may engage energy generation companies, often called Independent Power Producers (IPPs) or Energy Service companies (ESCOs) to supply energy for terms of twenty years are more. These companies are engaged to engineer, build, operate and maintain a renewable utility that blends renewable energy sources from onsite and sometimes off-site locations.

Adding renewable energy to a community requires a fairly large-scale investment. However, there are an increased number of active sources of capital investment to back Power Purchase Agreements (PPAs). PPA agreements bind a place to its energy source and can contain very specific information regarding the scope of the project, the payments, the reliability, the repair service levels, warrantees, legal, cost escalation and other matters. These agreements fall within a specialized area of law that varies in every state based on the regulations made by the state's public utility commission and federal regulations.

The costs of energy efficiency and renewable energy equipment for an individual property can be rolled into a residential or commercial mortgage or lease, financed through a bank as part of an energy efficiency upgrade, paid for via a flat monthly charge to a housing authority or financed through a third-party ownership company, among others. The best choice will depend on what options are available in the region where the net-zero energy project is located.

Residents of ZEC homes, buildings and communities utilize less energy because net-zero energy projects employ energy efficiency measures. The initial cost of installing energy generating equipment requires some combination of up-front investment and/or financing to fund required equipment and services. The key goals of the ZEC planning team should be to answer the following questions:

- How can the overall cost and risk be allocated among the consumers, builders and community?
- How can the consumer cost be held to power grid-parity or less, and improve thereafter?
- What entity will be formed or tasked to manage shared investment and energy billing aggregation for the ZEC?
- Regarding the ZEC financial proposition for new homebuyers:
- How do the overall costs for equivalent non-ZEC and ZEC homes compare on a twenty-year basis?
- What is the status-quo non-ZEC home to compare the ZEC home to?
- What would be the average home size?
- What does a non-zero energy house cost?
- What would the 20-year cost of energy be for a non-zero energy home? (include: www.eis.gov projected energy cost increases)
- What are other operational energy costs for energy across 20 years (maintenance, consumables, equipment replacement)?
- What is the cost of a comparable average-size zero energy ZEC home?
- What are added costs for energy efficiency of a ZEC house?
- What are added costs for energy-efficient appliances?
- What are added costs for geothermal heat, cooling and hot water?
- What are added costs for solar or wind equipment?
- What are other operational energy costs for energy across 20 years?

- What are the costs for finance charges and interest to finance all of the above in the mortgage?
- What are the value of all incentives, tax-credits, and rebates for the ZEC home?
- What is the projected increase in market value over the average non-ZEC house?

*Conclusion*

Those involved in the financial analysis for a proposed ZEC need to analyze and develop financial models that compare the cost/benefits of a ZEC development with the status quo development. To be better, a ZEC must evolve unique economics in the property life-cycle that are conveyed to a variety of constituents in a compelling way.

# SECTION 4
# TWENTY-FIVE OBJECTIONS TO A ZEC

Sometimes people express beliefs that are counter to the inevitable need to achieve a sustainable energy solution. Below are a few of the ideas that ZEC planners might encounter:

(1.) Many people do not understand the dramatic improvements in solar panels over the past decade. Likewise, they may not know about the technology improvements and lowered costs. This can cause hesitation in people, making them feel they want to wait until there is a larger mass adoption of the renewable energy technology and ZECs.

(2.) Because of the trend in cost reductions, some people may think ZEC planners would be best to wait until prices come down further.

(3.) Some people simply do not want to change. Change is difficult even for those who are willing, and some others may resist changing anything at all.

(4.) Can we determine who has the duty to create the ZEC? Why is this not the responsibility of the utility company, or the city, or the housing department? Someone should have already done something to make our power green. Find out who they are—and fire them.

(5.) America is rich with natural gas. We can just burn gas here at home for the foreseeable future. Possibly, by then, we will have cold-fusion energy.

(6.) Solar panels are an eyesore. Wind towers make soup out of birds. Hydroelectric creates reservoirs that are also good for fishing and boating—that is the way to go.

(7.) There is nothing wrong with hydrocarbons, and we have an unending supply. This is all a ploy by some people who wear Birkenstock shoes. They want our oil company profits to be paid as taxes and then re-distributed as subsidies to buy solar panels for poor people.

(8.) There is no proof that there is climate change. Global warming is being revised to global cooling. Greenhouse gases are a natural occurrence.

(9.) I need to be assured that this Smart Grid will not fight with me when I am turning things on and off.

(10.) I do not need an electric car. Those are like tin cans and my truck runs fine on gasoline. The return on the investment is just not worth it.

(11.) We do not want to be first.

(12.) The up-front cost of sustainable building programs will make property prices too high and the return-on-investment (ROI) period is too long.

(13.) Isn't the utility company solving this problem already?

(14.) The use of government incentives is simply unacceptable to me.

(15.) Investments in energy efficiency and renewables may not be reflected in real estate appraisals.

(16.) The technologies are not proven.

(17.) Making America competitive means we cannot support higher standards.

(18.) This is not our core business and therefore we should not get involved with it.

(19.)  We do not have the budget...we simply cannot afford it.

(20.)  We cannot afford a "green" premium.

(21.)  The payback needs to be less than two years.

(22.)  How do we know if the solar energy will last?

(23.)  It is too expensive, it costs too much, and our town is broke.

(24.)  Homebuyers are more interested in kitchens and man-caves; we cannot tell them they have to buy solar panels. It is their house.

(25.)  I just do not see that we have to change anything.

### Conclusion

Stagnant assumptions about the environment, economy, cost and efficiency of renewable energies and of super-insulated buildings cloud people's views about the possibilities for a more immediate change. There are also political ideologies that affect people's perceptions and willingness, or unwillingness, to change.

There are practical, economic, aesthetic and social benefits to living in a ZEC, but other people's concerns are real, as well. ZEC planners can overcome opposition when they demonstrate respect by meeting opponents or doubters where they are.

Notes

# SECTION 5
# HOW WE USE ENERGY

*Transportation: the largest portion of energy consumption in the US, accounting for twenty-seven quadrillion (27 QUAD) British Thermal Unit (BTU) of energy each year. Residential and commercial buildings, a close second at twenty quadrillion (20 QUAD) BTU each year, nearly one-third of all energy. Industrial energy is the third highest use of energy, consuming an additional fourteen quadrillion (14 QUAD) BTU of energy each year. (US Energy Information Administration 2013)*

Implementation of building energy efficiency, along with transitioning to renewable energy sources as in zero energy communities (ZECs), could make a very material contribution to the realization of effective US energy policy. In turn, reducing energy consumption could create more significant benefits to the national economy, the environment, and individual prosperity.

### The Present and the Energy Future

According to the Rocky Mountain Institute, by 2050, the US can phase out its use of oil, coal, and nuclear energy by using energy more efficiently and relying on natural gas and renewables to fuel the US economy. The energy efficiency opportunity accounts for more than half of the business-as-usual consumption in 2050 (assuming frozen efficiency from 2010–50).

Aggressively exploiting this energy efficiency and renewable energy opportunity makes the transition from oil and coal cost-effective, and enables a roughly one-third reduction in natural gas consumption and a major investment in renewable energy.

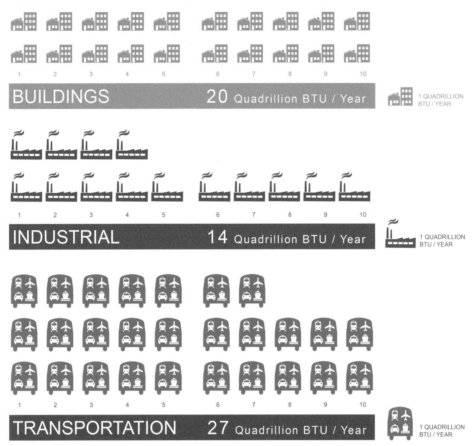

*Figure 5 - US ENERGY CONSUMPTION FOR 2010 USING DATA FROM REINVENTING FIRE. ZERO ENERGY COMMUNITIES REPOWER COMMUNITY-BASED TRANSPORTATION AND BUILDINGS WITH RENEWABLE ENERGY. THIS ACCOUNTS FOR MOST OF THE ENERGY, AND MOST OF THE ENVIRONMENTAL AND ENERGY SECURITY PROBLEMS. ENERGY EFFICIENCY IS AN IMPORTANT COMPONENT OF THE ZEC. (Lovins, Reinventing Fire: Bold Business Solutions for the New Energy Era, 2011)*

The organizing theme for resolving energy efficiency and a transition to non-carbon fuels is referred to as "de-fossilization," according to the Rocky Mountain Institute.

### Use of Energy for Transportation

As stated and diagrammed, our largest use of energy is for transportation, powering the trains, planes, cars, trucks and buses that move people and materials. There are significant initiatives underway aimed at curbing the amount of non-renewable energy utilized for transportation: reduction of transportation requirements and adoption of mass transit systems, fuel efficiency and alternative fuel for vehicles.

ZECs materially affect transportation energy use because they accommodate mass transit, improve the walkability of neighborhoods, and support electric vehicles that can be operated with renewable energy.

### Use of Energy for Buildings

The second largest consumer of energy is the operation of buildings and their internal systems. These functions include, but are not limited to: lighting, heating, ventilation, air conditioning, refrigeration, elevators, information systems, streetlamps, water, irrigation, wastewater, telecommunications, financial systems and transportation management. Much of that energy production creates pollution and is lost to inefficiency. Almost all buildings require a constant supply of energy to support their operating systems and the uses of energy by their occupants. Lighting the way for a growing world involves massive and growing consumption of energy.

ZECs provide an alternative to the growing demand for energy used for powering our buildings, and unlike a home that is a one-off zero energy home, ZECs provide the potential for a very material reduction in energy consumption. ZEC development methodology needs to be adopted broadly and improved continuously.

ZEC planners can develop design guidelines to prescribe required energy conservation and renewable energy technologies in the ZEC. Specific technical requirements for buildings are determined after understanding the baseline requirements (those required to meet building codes) and exploring the cost implications of the ZEC energy system infrastructure requirements.

### Use of Energy for Industry

"Embodied energy" refers to energy that has previously been expended to extract, manufacture or transport, and situate a material, equipment or a structure. This energy is consumed, never to be recovered, and is therefore embodied energy contained within these goods, products, food, and materials that we use in all aspects of our lives.

By reducing our consumption and avoiding waste, we reduce the use of energy otherwise required to make and deliver the goods for our use. In addition to curtailing the consumption and waste of goods, we can source goods locally, thereby reducing the embodied energy demands through the process of minimizing transportation and delivery.

A ZEC can provide a platform for greatly reducing use and waste of embodied energy. This can be done by first, addressing the energy embodied within building materials, and secondly, by educating those in the community about effective ways to reduce usage and waste of goods, especially those that require extraction, or non-renewable energy to manufacture and transport to the consumer.

The United States Green Building Program, LEED Green Building Rating System, and LEED for Neighborhood Development programs incorporate methodologies that can significantly reduce the embodied energy components of a ZEC and are highly recommended companion programs for ZEC planners, as the programs align with ZEC energy goals. (See Section 14: Other Programs and Standards)

### ZECs Broadly Reduce Our Use of Energy

A ZEC contributes to the reduction of non-renewable energy through improving energy performance in transportation, buildings, and embodied industrial energy. By establishing appropriate designs and programs of training and performance measurement, a ZEC can make a difference in our energy utilization, support the transition to renewable energy from fossil fuels, and provide a sustaining program for measurement and continuous improvement—toward the ultimate goal of helping any group to achieve a net-zero energy balance.

The foundation of a ZEC is conservation of energy. Conservation is achieved through both the application of technology and the modification of occupant behaviors. There are philosophical, environmental and economic advantages to conservation.

### Conservation of Land

In the US, it is customary to build altogether new places upon Greenfield land sites. It is also possible to leave the natural resources of those Greenfield sites intact, and instead, redevelop and restore an existing site. A site that has been built on previously is called a "Brownfield" site.

From the standpoint of conservation, building upon, or renewing a Brownfield site is far more advantageous to environmental preservation. It is possible to utilize both Greenfield and Brownfield sites in a ZEC project. As an example, the solar garden (a field of solar panels) may be installed on top of an environmentally hazardous site to provide energy to a nearby Greenfield ZEC. Conversely, a historic neighborhood could renovate every building and use parkway green spaces (Greenfield) for producing solar or wind energy for the ZEC.

ZEC planners face different challenges. A Greenfield ZEC planning group can lay out lot lines, streets, and plant trees in ways that can optimize solar energy generation. A planning group that is making a one-hundred year-old neighborhood (Brownfield) into a ZEC does not have those same opportunities; but instead, has a strong community fabric in place to provide leaders, sponsors, and skilled volunteers the opportunity for collaboration and involvement.

### Energy Efficiency

Achieving a balance of community supply and demand for energy starts with conservation and energy efficiency, which reduces energy demand by the most efficient and least expensive means.

Reduce the demand for energy to the lowest point practical, and then consider the requirement for renewable energy. The science for achieving conservation of energy is called "energy efficiency." Through advances in equipment, appliances and building technologies, it is a burgeoning field.

Transportation, buildings and industry consume the greatest amount of energy, but none of these three is the single largest use of US energy. Within the three areas of consumption lies a single subset of energy consumption that is larger:

> *What is the single largest consumer of energy in America? It is also responsible for the most CO2, and is responsible for the most imported oil. Waste heat wins all three races, a triple crown. (Sherwin 2010)*

In a ZEC, conservation is the most important goal and a key determining factor regarding the quantity of renewable energy required to balance supply and demand of a net-zero energy balance. It is necessary to determine how much energy use can be avoided before understanding the ZEC energy demand, and the balance of on-site renewable energy that is required to net that out.

The "negawatt" is often the least expensive and most useful energy source because it negates the use of energy. Government programs at the federal, state, county and local levels all participate in programs aimed at creating energy efficiency. Whether for energy efficiency research, technology commercialization, business or consumer information, or financial incentives or de-incentives, these programs stand to reduce electrical energy consumption by one-half or more.

> *A "negawatt" is a theoretical unit of power, representing an amount of energy saved through efficiency, measured in watts. (Lovins, Chief Scientist and Founder, Rocky Mountain Institute, 2013)*

The National Laboratories, particularly the National Renewable Energy Laboratory (NREL), focuses research on efficiency. Utility companies also provide information and incentives for weatherizing buildings and replacing inefficient equipment and appliances, in addition to shifting the timing of energy demand.

### Timing of Our Demand for Energy

The ZEC electrical system must provide for all of the community's needs during times of peak electrical use. The peak usage can be met by the energy generation capabilities of the ZEC or by an interconnected regional utility system. When the energy supplies are combined, they must serve that load by providing electricity. The amount of "peak load" determines the sizing of transmission lines, substations, transformers and utility distribution power lines, and has significant impact on the capital cost of the electrical system. The entity ultimately responsible for providing enough electricity is said to be the provider of "power-of-last-resort." This responsibility falls on the regional grid when a ZEC is interconnected with that grid.

Any means used to reduce the peak load and normalize the requirements for power produces energy system-efficiency and cost-efficiency. This peak-shaving is not in itself energy efficiency, but peak-shaving can be accomplished through energy efficiency projects, producing great value.

Here are examples of the type of energy efficiency projects that may or may not affect peak load:

(1.) On a hot afternoon during the summer season, many houses in a ZEC neighborhood may be utilizing conventional air conditioning systems that create a peak-load condition for the ZEC. If the ZEC had used high-efficiency geo-thermal heat pumps for air conditioning systems, their energy utilization, and the peak load, could have been reduced.

(2.) Neighbors in a ZEC that utilize extensive street lighting enjoy the sense of safety from the well-lit streets and sidewalks. An energy efficiency proposal to turn those lights off from midnight to 4:00 am each night achieves energy efficiency, but likely has no effect on the peak load of the ZEC. Both are valuable outcomes, but an energy efficiency initiative that reduces peak load is probably more valuable.

Equalizing energy demand across the daily clock can be achieved by changing the routines in our energy use. Once people understand the context and concepts of energy conservation, they can contribute material value through small adjustments in their behaviors, like running the dishwasher at night instead of during peak loads. One of the most compelling aspects of ZECs is the potential for unilateral behavioral modification. Asset utilization efficiency is difficult to achieve when the load requirement is modulating continuously. Technologies also enable consumers and businesses to automate the timing of energy use; e.g., starting the dishwasher can be pre-programmed to automate the delay until later in the night when less energy is usually being used for lighting and air-conditioning.

Utility companies are experimenting with smart phone applications that notify you when there are peak loads stressing the grid or if your monthly energy bill is exceeding your average or established budget target. ZECs can adopt the same systems and also provide metrics on the over-arching community goals for energy use.

*Transportation, building energy, and energy technology form a complex dilemma that is changing. Alison Wise, Principal, Wise Strategies, an energy strategy and renewable energy expert, recommends the following for ZEC planners: (Wise 2013)*

- *Learn the current information about transportation solutions, building technologies, energy efficiency, renewable energy, and energy control technologies.*
- *Innovate in the area of finance to improve cost/benefit and to balance immediate and long-term outcomes. Explore grants, crowd sourcing, and community funding.*
- *Build community sponsors, experts, local stakeholders, initiators, influencers and storytellers who communicate through PR, social media, events, news and education.*

*Figure 6 - TESLA MODEL S: A SLEEK, ALL-ELECTRIC VEHICLE FEATURING HIGH-END LUXURY, STATE-OF-THE-ART DESIGN AND AN ALL-ELECTRIC DRIVING RANGE OF 300 MILES.*

Conservation of energy by transportation is relevant to ZECs in three ways:

- ZECs are typically planned to accommodate mass-transit.
- ZECs are planned to be walkable neighborhoods, reducing the use of automobiles.
- ZECs are required to accommodate charging of electric vehicles.

### Buildings' Energy Efficiency

Energy conservation is the foundational requirement of a ZEC initiative. Before attempting to balance the sources and uses of energy, the energy requirement should be reduced to the lowest practical minimum. As will be described in detail later in this guide, conservation of energy use is achieved by:

(1.) Improving the insulating properties of buildings

(2.) Equipping buildings with energy-efficient systems for lighting, heating and cooling

(3.) Adjusting the orientation of planned new buildings relative to solar access and shading from trees

(4.) Adjusting the behaviors of the occupants of the community

(5.) Accommodating efficient means of transportation to, from, and through the community (mass transit / electric vehicles)

### Building Energy Conservation through Technology

Eliminating wasted energy is achieved through better building technologies that include insulation systems, solar orientation, natural day-lighting and efficient heaters, air-conditioners, water heaters, appliances, lighting and lighting controls.

> *A ZEC project requires both energy conservation and renewable energy sources. First, energy efficiency is optimized by choosing highly efficient buildings, appliances, and electronics. Second, it is realized further by conservation awareness leading to effective conservation practices, according to NREL. (Carlisle, AIA, VanGeet and Pless 2009)*

The ongoing discussion on USGBC LEED, HERS, and Envision suggest that the conservation standards for buildings are close to becoming routine. It is crucial to reduce energy use as much as possible before looking to install onsite or offsite renewable energy.

### Building Energy Conservation through Modifying Occupant Behaviors

The technological approaches to conservation are complex, but may be easier to achieve than modifying behaviors. The group orientation of a ZEC opens the door to conservation by means of modifying behaviors. The challenge is in identifying and agreeing on where the line is drawn to mark where a behavior is too wasteful. Each person's minimum standard of convenience is different. ZECs seek to strike the balance within personal extremes.

Conservation in ZEC planning that extends to modifying occupant behaviors by reducing the waste of energy is achieved through modifying behaviors such as keeping doors and windows closed while heating or cooling the building, turning off unused lighting and appliances, dressing in order to moderate the use of heating and cooling, and eliminating trickle-current plug loads for various powered equipment such as cell phone chargers.

It is also advisable to reduce the use of energy at times of the day when overall demands are highest, such as using a whirlpool hot tub and sauna bath only at night.

Daniel D. Chiras, in his book *The Homeowner's Guide to Renewable Energy: Achieving Energy Independence through Solar, Wind, Biomass and Hydropower* (Chiras 2006), explains how his family built a net-zero energy home, and how his family made adjustments in their routines in order to achieve balance. Some of the suggestions made in the book are simple to implement and save considerable amounts of energy without significant expenditure or sacrifice.

To give a personal example, adjusting the temperature to 110°F and adding an insulating cover on my own old and inefficient home hot water heater has reduced my bill, increased the amount of hot water available when needed, and left me feeling true to the goal of energy efficiency.

*You can suggest that people wear warm slippers in winter and turn their heat down to conserve energy. You cannot, however, make a law requiring everyone to own warm slippers. You have to bring people around by appealing to their intelligence and will. (Tinianow 2013)*

### Industry Energy Efficiency

Use of energy by industry: The third largest consumer of energy is primary energy used to produce the supplies needed to enjoy life. From food to soap, cars to cardboard and underwear to ball bearings, every product used has been manufactured using energy. As I explained in Section 5, How We Use Energy, the energy used to manufacture and deliver products and supplies is referred to as "embodied energy."

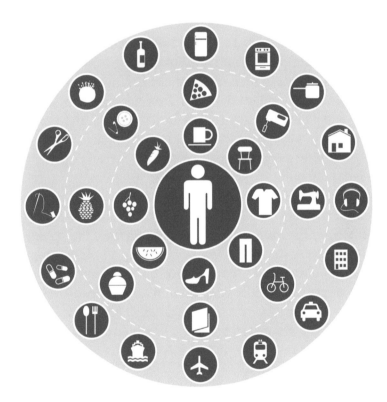

*Figure 7 -ENERGY IS EMBODIED IN EVERYTHING THAT WE DO IN LIFE. THE "EMBODIED ENERGY" CONTAINED IN OUR FOOD AND OUR POSSESSIONS IS SIGNIFICANT.*

Saving material saves industrial energy. America produces a large share of energy in order to produce the many materials we use for food, shelter, transportation, communication, and everything else we use in our lives. All of the goods that people own contain energy embodied in the manufacturing, delivery, installation and operational processes connected with these goods.

*Even the disposal or reclaiming processes required to re-purpose the stuff as trash, or as new feedstock, could consume industrial energy. Even passive activities like using a smart phone incorporates embodied energy. Every time you dial, text, email, or perform a web search, data centers respond by processing data. Data equals energy. The ones and zeroes correlate to electrons flowing through computer circuits. It may surprise readers to know that this is one of the fastest growing uses of electricity in the world. (Whitcomb, Digital Energy 2009)*

ZEC education can help people understand the need to moderate their uses of all goods and can help them understand how to select goods with better life-cycle impacts. Goods are environmentally advantageous when they are created using less energy, delivered using less energy and operated using less energy. Additionally, products that can be re-purposed or bio-degrade after serving their lifecycle also contribute to the environment. Energy-efficient products can also save costs to the consumers.

The green building standards applied in ZECs have already provided a terrific applied-work project example from which the ZEC team can learn. Industrial material resource improvements related to building materials, furniture, fixtures and equipment are managed within the USGBC LEED rating system for neighborhood design and buildings.

The Earth's elements are valuable in their native states and in compounds. The "cradle-to-cradle" theory suggests we can manage the naturally occurring elements and chemical compounds, preventing exposure that endangers living organisms and preserving valuable materials for intergenerational use. For example, the mercury contained in old energy-efficient compact fluorescent light bulbs can be isolated from direct human contact, upcycled, and then used to manufacture new light bulbs.

*"Upcycling" is the term used to describe products designed with the recapture and use of natural and technical nutrients in mind. We might be designing a car for disassembly, so the steel, plastic, and other technical nutrients can once again be available to industry. We might be encoding all the information about all the ingredients and materials themselves in a kind of upcycling passport that identifies all the substances used in construction and indicates which are viable for future nutrient use and in which cycle. (McDonough and Braungart 2002)*

If we can all use less stuff, use more local sourcing, use environmentally improved products, and see that the stuff is properly recycled when we are through with it, we will conserve industrial energy, as well as other natural resources.

A ZEC provides a structure for a community to learn about, and practice, the conservation of all sorts of energy. A ZEC is big enough to make a difference, small enough to be available, and potentially enough to make a difference for a whole community.

*Standing in the way of change are corporations who want to continue worldwide deforestation and build power plants, who see the storage or dumping of billions of tons of waste as a plausible strategy for the future, who imagine a world of industrial farms sustained by chemical feed-stocks. (Hawken 1993)*

# SECTION 7

# UNDERSTANDING THE ELECTRIC GRID

A basic understanding of the electric grid is helpful to those interested in zero energy communities because a ZEC community contains what is essentially its own electric microgrid, a community-level electric system that may have, or has the potential to have, intelligence built in at the local level that is similar to, and may be conversant with, the national grid.

>  *An electrical grid is an interconnected network for delivering electricity from suppliers to consumers. It consists of generating stations that produce electrical power, high-voltage transmission lines that carry power from distant sources to demand centers, and distribution lines that connect individual customers. (Kaplan, et al. 2009)*

Since its inception in the Industrial Age, the electrical grid has evolved from an insular system that serviced a particular geographic area to a wider, expansive network that incorporates multiple areas. Historically, all energy was produced near the device or service requiring that energy. Electric grids in the US were regulated in order to serve the load, supplying electricity to reliably meet the demands of all, a program that would enrich the nation through electrification.

>  *Electric systems spread across the country, even to locations where extending power was not viable economically. The grid evolved through the construction of large base-load power plants, including hydroelectric, coal and oil powered plants, transmission circuits and distribution systems. Eventually, in 1958, America's first nuclear power plant opened. (Parker and Holt 2007)*

The average electrical demand of a community is referred to as the load or "base load." The maximum load is referred to as the "peak load." As electric demand increases or decreases throughout the day, (and it can vary considerably) the average load is usually less than 50% of the peak load. Power plants will ramp power production up or down to maintain a balance between the supply of electricity available on the grid and the demand. The activation of generation assets is referred to as "dispatching" power.

During the 20th century, the institutional arrangement of electric utilities changed. At the turn of the century, electric utilities were isolated systems without connection to other utilities and serviced a specific service territory. In the 1920s, utilities merged, establishing a wider utility grid as joint-operations saw the benefits of sharing peak load coverage and backup power. In addition, electric utilities were easily financed by Wall Street private investors who backed many such ventures. In 1934, with the passage of the Public Utility Holding Company Act (USA), electric utilities were recognized as public goods of importance along with gas, water and telephone companies. Utility companies were given outlined restrictions and regulatory oversight of their operations.

As the 21st century progressed, the electric utility industry sought to take advantage of novel approaches to meet growing energy demand. Utilities came under pressure from customers and regulators to evolve their classic topologies to better accommodate "distributed energy generation" (DEG), which takes power from local solar, wind and hydroelectric generators. DEG is a completely opposite approach than is the central-plant generator model that was the core topology of the US power grid as it was conceived. As distributed generation becomes more common, the differences between distribution and transmission grids will begin to blur and a major Smart Grid will emerge.

Electricity derived from solar and wind is not currently able to be dispatched. The weather determines wind speed, sunshine and cloudiness. When the wind is not blowing, turbines do not turn; when the sun is not shining, solar panels do not produce electricity. For this reason, wind and solar, when added to the grid, can be somewhat unpredictable, uncontrollable and intermittent. These unpredictable energy sources pose an additional challenge to grid management.

"Demand-side management" (DSM) is a grid management technique where retail or wholesale customers are requested, either electronically or manually, to reduce their load by voluntary energy reduction. DSM gives energy users the ability to control equipment during peak load conditions. Users can manually or automatically de-energize equipment like refrigerators or air conditioners, to reduce peak energy demand. By de-energizing such equipment for short intervals of time, the grid operator can materially reduce the amount of the peak load. Currently, transmission grid operators also use DSM response to request load reduction from major energy users such as industrial plants (Demand Response Symposium: industry action plan at PJM Interconnect 2008). It can be a cost-effective means of balancing supply and demand in the electric grid because it does not require additional fuel or equipment.

Another important aspect of demand-side management (also called demand-response or DR) is the method used by utilities and grid operators to dispatch calls for electrical generation to increase the electrical supply. Many buildings and campuses are equipped with emergency backup generators (e.g., airports, police and fire stations, hospitals, military bases, schools, resorts, data centers) that may be remotely dispatched when additional supplies of electricity are required.

The amount of voltage on lines in the electric grid must be tightly regulated. A few volts too high or too low for too long can cause transformers, capacitors and other equipment bringing electricity to where it is used, to fail, causing large blackouts until the equipment can be replaced.

Voltage spikes can also damage appliances and equipment connected to the grid. The required balance is primarily achieved by "dispatching" power generation from base-load power plants and "peaker" power plants. Peaker power plants are brought online as needed, but often cost more to operate than base-load plants.

Voltage on the grid is regulated by electric control centers operated by electric utility companies. The operators anticipate demands and adjust generation in order to provide the precise amount of required electricity at every given moment.

Electrical control centers are deploying advanced software to manage the balancing act between the increasingly complex demands of the grid, which includes conventional base load generation, peaker power plants, solar, wind power, energy storage, and consumer behaviors.

Electric companies, utilities and project developers are experimenting with large-scale batteries and other energy storage devices to help smooth the effect of intermittent renewable energy. However, these technologies are prohibitively expensive today.

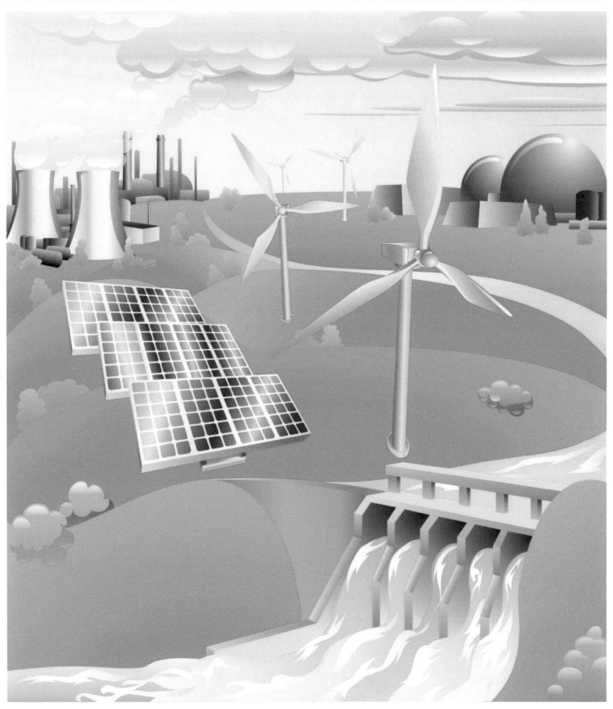

*Figure 8 – US UTILITY COMPANIES MANAGE AN EVER-BROADENING ARRAY OF NUCLEAR, COAL, GAS, HYDRO, WIND, SOLAR AND GEO-THERMAL POWER GENERATION ASSETS AS WELL AS DEMAND RESPONSE SYSTEMS AND ENERGY STORAGE.  OPTIMIZATION INVOLVES MEETING LOAD OBLIGATIONS, ENSURING RELIABILITY, MANAGING EMISSIONS AND PROFIT IN MARKETS THAT TRADE POWER AND CARBON CREDITS.*

*Balancing electrical supply and demand is a critical requirement of all electric grid systems. Public and personnel safety, utility operating economics, power reliability, coordination of independent operators and consumer convenience all depend on that balance. Operators, traders and dispatchers who work in control centers participate in a dynamic environment in which they can face episodes of chaos, and emergencies, when the balance of such systems fail. For example, problems at a FirstEnergy Corporation control center in August of 2003 caused a widespread blackout of the northeast US and Canada that affected fifty-five million people. (US/Canada Task Force - Power System Outage 2004)*

Evolving utility operating protocols, new software and improved technologies are constantly being developed to help improve the cost, quality and reliability of grid operations. Generating assets that can perform "firming" of power are being deployed in grid-scale applications.

This makes it easier to firm power, which is to balance the effects of intermittent wind and solar energy (Undisclosed 2013). It is conceivable that a ZEC of significant size, or a ZEC that is off-grid, would incorporate similar control and monitoring equipment to a large utility company. Micro-grid management of the ZEC could interact with the primary regional grid control, so that the perturbations in voltage between the ZEC grid and regional power grid would interact with one another in a useful way. The good news for ZEC planners is that ZECs connected with the power grid only have to balance supply and demand on an annual basis, and that any minute-to-minute imbalance is shifted to the utility that operates the regional power grid.

An off-grid ZEC requires energy generation capacity to meet the demand of the peak load, instead of the average load, and this would greatly increase the number of solar panels, wind turbines or other non-renewable power sources dedicated to the ZEC. This extra capacity to generate energy for peak loads would imply an electrical system management requirement on the ZEC that would greatly increase the cost, complexity and operating responsibilities.

### The Utility Business

Collectively, individuals and businesses in the US pay energy utility companies billions of dollars daily to power transportation, buildings and industrial processes like manufacturing. In turn, these companies spend billions of dollars on mining and extraction enterprises that provide coal, gas and other fuels. The transport of these resources is significant. Railroads and pipelines haul these fossil fuels to power plants where electricity is generated. These powerful power, coal, pipeline and natural gas industries are entrenched and often monopolistic industries that have, for over one hundred years, continued to accumulate massive wealth.

Are most of these large, embedded industries with growing profits motivated to implement renewable energy and energy efficiency projects that eliminate or reduce the need for their products? Not likely. Smart Grids and ZECs can minimize reliance on utility companies in a number of ways:

- Energy-efficient users, buildings and vehicles need less energy.
- Communities that utilize locally-sourced renewable energy do not purchase as much power or natural gas from the local utility.
- Zero energy communities that are on-grid require the utility companies' services to balance the short-term imbalance of energy supply and demand.
- Zero energy communities that achieve a zero-net balance do not compensate a utility in any way whatsoever.

ZECs have not been supported by US utility companies to date. The City of Boulder, Colorado entered into a partnership with Xcel Energy to develop a Smart Grid infrastructure and increase renewable energy utilization within the city. Boulder viewed the project as a colossal failure, and subsequently passed a ballot proposition that provided the funding to develop a municipal power utility. Kelly Crandall, a Sustainability Specialist, City of Builder, Colorado says:

> We are concerned with sustainability, and having self-control in regard to our energy. When I think of net-zero energy, I think of small neighborhoods where microgrids use renewable energy balanced with small-scale storage. I wonder if anybody is thinking about how to implement this in eco district terms... where it is a zero-energy community and is also incorporating other environmental services - like transportation. [The ZEC idea of] combining net-metering and distributed energy and transportation is compelling. (Crandall 2013)

> Most utilities are investor-owned utilities (IOUs) that have a monopoly within a region and are for-profit entities, although some are co-ops or municipally owned. Utilities are regulated by both federal and state public utility commissions that must balance public/private interest, but ultimately guarantee profitability for the utilities. Building ZECs could end or reduce the lucrative profits such companies realize today and challenge existing regulations and the status quo of these companies' practices of the last 100 years. To put the size of this industry in perspective, the top ten electric utilities (by customers) enjoyed gross revenues totaled $69 billion dollars during 2011. (US Energy Information Administration 2013)

If utility companies would develop new services for ZECs and new business models that compensate them for the value they provide, ZECs could be a lucrative aspect of their businesses.

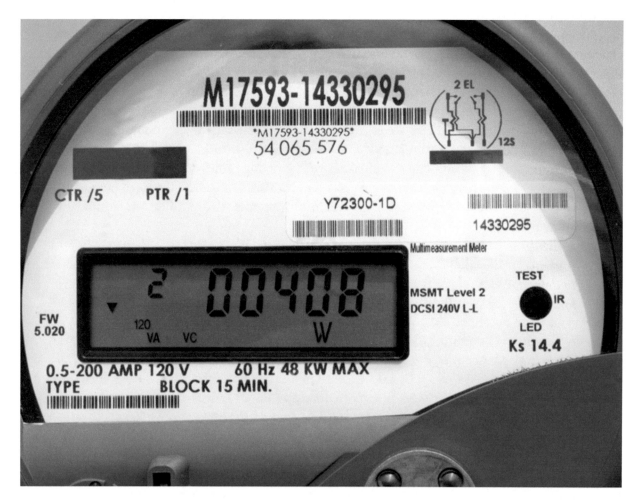

*Figure 9 - A SMART METER RECORDS CONSUMPTION OF ELECTRIC ENERGY AND COMMUNICATES THAT INFORMATION BETWEEN THE UTILITY COMPANY AND THE ENERGY CONSUMER AT INTERVALS OF ONE-HOUR OR LESS. THIS TECHNOLOGY ENABLES TIME-OF-USE RATE STRUCTURES, AUTOMATIC POWER OUTAGE REPORTING, ELIMINATES MANUAL METER-READING AND PROVIDES CONSUMERS WITH USAGE DATA.*

*Utility companies could embrace ZECs with strategic changes wherein they take ownership of financial or operational stakes in ZECs. Thus far, there is little evidence of that happening. By contrast, utilities in twenty-nine states are now required or encouraged to utilize renewable energy sources through "renewable portfolio standards" (RPS). (DuVivierr 2011)*

The RPS mandates that utilities will incorporate renewable energy from specific sources (e.g., 10% wind energy plus 10% solar energy) and establish deadlines and penalties for failing to comply. Could a ZEC foster a shift that creates a material, bilateral benefit for utilities and the communities in which they serve? The City of Boulder, Colorado is in such a discussion with their utility today.

Utility companies and cooperatives operate in a high-stakes arena where new technology, large capital expenditures, long-term project financing (usually thirty-years), uncertain regulatory changes, requirements to arrange capital financing and to purchase fuels (e.g., uranium, coal or natural gas) within volatile markets, all present significant risks. These utility companies vet the business, regulatory and engineering issues, and make sophisticated bets, funding programs that cost tens- to hundreds-of-millions (or even billions) of dollars.

The regulated model guarantees utility companies exclusive operating territories and profitable operation. State public utility commissions (PUCs) provide governance of utility programs, setting the retail rates we pay for power. Federal regulations also impact utilities, particularly power plant environmental, inter-state operation, and interoperability mandates. Now, and for twenty-five years previously, federal and state regulations have introduced greater competition to the utilities.

> *Traditionally, US power utilities' primary goals were to build out power infrastructure, to assure electricity for every American, to serve the load and to deliver at least 99.99% reliable electricity at all times. These goals are no longer plausible given the increased demand for power from all sectors. This is evidenced by the increasing growth of blackouts. These occurrences have increased in frequency, breadth and duration. Consider this: during 2011, the average American consumer went without power for 112 minutes, the highest duration of power outage in ten years, according to industry authority PA Consulting. At the same time, utility company spending for replacing local distribution equipment, measured by average cost per customer, rose from $163 to $232 according to Ventyx. (Jonathan Fahey 2013)*

The cost to maintain the grid is rising, but the cost to replace the aging power infrastructure would be enormous. Still, the occurrence of blackouts is increasing because old equipment is failing. These difficulties are captured by the two most significant problems facing today's utilities:

(1.)  America's aging power grid infrastructure

(2.)  The great risk that the aging grid poses to intentional attack to US homeland security

These problems are symptomatic of a much larger systemic challenge. In the decade since 9/11, no fundamental change in the American energy infrastructure has occurred. In fact, the aging grid infrastructure is becoming even more unreliable. Is it acceptable to simply say the problem is just too big to solve, especially when we can imagine the failure of the grid as a likely result of a terrorist attack? The consequences of such an attack in terms of collateral damage and long-term effect are potentially staggering. Such scenarios are in the current public discourse and provide added compelling and urgent motivation to create ZECs.

> *Several months after 9/11, the National Science Commission delivered a report to then-President George W. Bush, indicating that a widespread and long-term power outage was the highest risk of terrorist attack on American soil. (Committee on Science and Technology for Countering Terrorism, National Research Council of the National Academies 2002)*

*Game-Changing Factors in the Power Industry*

According to the findings of an Edison Institute Report released in January 2013, disruptive challenges are occurring in the electrical utility industry due to a convergence of factors (Kind 2013). These include:

- Energy efficiency (EE)
- Distributed energy resources (DER)
- Energy demand-side management (DSM)
- A continuum of renewable energy (RE) technology and market innovations becoming economically viable (e.g., solar photovoltaic)

Increased use of DER, which places power generation closer to the points of energy use, has already absorbed a 1% market-share of all retail power in the US. In addition, a regulatory environment that is biased toward encouraging competition, along with the impacts of RE, DSM, and EE, all serve as disruptive forces, forming a significant threat to the status-quo in the utility business sector.

> *The continuing rise in cost for conventional energy (16% of the US retail electricity market has a rate above $0.15 kWh) and the falling price for solar energy are creating a situation where DER market-share would quickly grow to 14%, and then grow exponentially. (US Energy Information Administration 2013)*

Developing a ZEC is complex, involving law, brand, business, finance, technology, trading and services. Utility companies have experts in all of those fields. In my opinion, there are no organizations better qualified to assess the technical and financial issues of a ZEC and to provide the necessary engineering, equipment, services, maintenance, power restoration, and customer service than America's power utilities.

However, the balance of energy sources and uses in the ZEC model challenges the foundation of traditional utility companies. ZEC planners need to be cognizant of the turmoil their local utility company may be experiencing. The utility companies may not understand their own futures clearly, making it difficult for them to commit their support to ZEC projects, even if their desire is to improve operations and delivery of energy.

> *Utility companies may bring great innovation to ZEC projects someday. The up-front financing requirements that ZEC developers face could reduce or disappear entirely if utilities got involved. Until such time as the utility companies have developed ZEC support programs and completed work with PUCs to establish rate structures and tariffs, ZEC planners should consider developing their ZEC independently from utility companies. (Lyng 2013)*

The more ZEC initiatives are underway, the more momentum to convince utilities to invest as partners. Jeff Lyng, Senior Policy Advisor, The Center for the New Energy Economy, Colorado State University, sees opportunity for utilities to adapt their business models to ZECs, where they are compensated for providing the services the community seeks:

The obstacle is not a matter of technology limitation, so how do planners bring the utility into the fold? Utility companies want to understand their customers. Perhaps they could be approached in the following way: "We as a community are very motivated toward energy efficiency and renewable energy, and we want to have a super-charged energy program. How might we work together?"

> *In addition, where the grid is stressed, where the risk of blackout is high, where there is a grid technical advantage to developing a ZEC that has a cost advantage to the utility and the ratepayers, the ZEC can solve real utility company problems. Engineering and political drivers can drive pilot ZECs in advantageous locations. What is most needed is policy and leadership, and courage. (Lyng 2013)*

The government loans, investments and grants aimed at advancing energy efficiency and renewable energy cause controversy and leave America divided. Where do you and the people you know stand on these issues of government energy policy?

> *A significant portion of the $831 billion dollar American Reinvestment and Recovery Act of 2008 (ARRA) was appropriated for energy advancement in the United States. (Patton Boggs L.L.P. 2009)*

Policies that inform our trade agreements, laws, tax treatment, and regulatory functions have implications for ZEC formation and operation, albeit none to date that name ZECs specifically. Still, current policies exist that regulate particular aspects of a ZEC and the associated markets for electricity, renewable energy and energy efficiency. Taken altogether, this area of policy is profoundly complex and inconsistent from region to region. Policy is made at the federal level, as well as at the unique state and municipal levels.

## Notes

# SECTION 8
# POLICY

ZECs must comply with regulations that may include design standards for building construction and particular requirements governing the sale of electricity. ZECs may have to gain approvals, and/or variances, that entitle their development to proceed, and that may potentially include public approval processes. Construction and occupancy permits are issued by the authorities with jurisdiction at the location of a ZEC, including local building departments, fire marshals, public works departments, tax administrations and public utility commissions. ZECs may also have to work with agencies to approve, or administer, financial incentives programs.

Trade agreements affect import and export, restrict subsidy programs, regulate global access to products and technologies, and establish tariffs, duties and taxation. All of the aforementioned directly influence equipment prices for RE/EE projects, electric vehicle costs and a host of other market factors.

> *For example, heavily subsidized Chinese solar panels compete with European Union (EU) and US solar manufacturers. The Chinese were so competitively aggressive that prices were reduced by 75% compared to US and EU manufacturers' prices, resulting in the failure of sixteen solar companies in the US and EU. The Obama Administration and the European Union have each decided to negotiate settlements with China in the world's largest anti-dumping and anti-subsidy cases involving China's roughly $30 billion-a-year in solar panel shipments to the West. Huge shipments from China have driven solar panel prices down by three-quarters in the last four years. (Bradsher 2013)*

Still other laws prevent export of high technology from the US to many countries in consideration of national security concerns. At the same time, programs of the US Department of Energy national laboratories assist local communities, the scientific community, utilities, manufacturers, and government, industrial and individual users of renewable energy.

The Kyoto Protocol, a global sustainability agreement, obliges over five hundred US cities be "green" in their conduct and place-making activities.

Each US state's unique electricity regulations control the terms of their utility companies' franchise agreements with municipalities and industrial and retail customers. These complex legal regulations may define the regulated electric rates, the acceptable level of reliability, the repair service levels required, and permissible interconnections to the grid, on-site generation, power resale, and provide (or not) for net-metering, and virtual net-metering solar gardens—all according to interpretations.

There is a constant churn of policy mandates that requires ZEC teams to have peripheral vision and expert counsel, plus a process for maintaining inspection into every aspect that could provide obstacles or opportunities to the ZEC.

### Regulatory Environment

Before the year 2000, regulations were in place that largely supported monopolistic and investor-owned, regulated public utility companies (IOUs) and rural electric cooperatives. These regulations made it hard for other groups within a service area to generate, distribute, or sell power. In current practice, particularly because of distributed energy generation (DEG), the realm of energy regulation is changing and increasingly complex.

> Newer regulation has created programs for solar net metering, community solar projects, energy districts, and geothermal utilities—making them more practical than ever before. (DuVivierr 2011)

Other regulations have changed laws affecting the wheeling of power (moving electricity over another's transmission lines/system), net, formation of utility districts, and the use and operation of solar panels, wind turbines and geothermal systems.

To avoid legal obstacles ZEC planners and developers are advised to seek expert counsel, seek to understand the laws that apply in their jurisdiction, and keep abreast of the sometimes fast-changing legislation.

New regulations have emerged, in part, because renewable energy technologies and energy-efficiency businesses have grown so much in recent years. What is less known in the public sector are the changes that have come as a result of active development of ZECs by the federal government. For instance, the Department of Defense (DOD) and Government Services Administration (GSA) are quickly retrofitting and building new energy-efficient buildings, setting up off-grid ZECs for military bases, and taking other measures to avoid reliance on foreign energy sources.

> According to the Assistant Secretary of the Army, the Army's expenditure for energy exceeded one-billion in 2011, and that energy supply is strategic. (Hammack 2010)
>
> The most comprehensive guide to date covering US electrical power is the Renewable Energy Reader. (DuVivierr 2011)

This resource provides background and comparison of local ordinances addressing solar rights, state regulations (including RPS mandates by state), federal standards, and information about issues of social justice related to energy.

As public interest continues to rise, whether related to broad concepts of law or the details of solar installation aesthetics, control of vegetation, solar access, and many other property rights issues, this resource explains the related policy, laws, local ordinances and other requirements.

ZECs' ability to supplement their onsite generation capabilities with off-site renewable energy from a solar, wind, or other renewable energy provider depends on the government regulations in effect at a particular jurisdiction. This has much to do with the unique laws of each state that establishes provisions for the wheeling of power over transmission lines. Another method that can be used to connect offsite renewable power to a ZEC is called "virtual net-metering," a method of accounting for the net-balance of electricity sources and uses located at different physical locations and connected to different electrical meters.

*NOTE:* Whenever considering offsite energy for a ZEC, one must consult an expert in local state energy laws and policy. Some federal (FERC) regulations make exceptions for military facilities, and sometimes state utility commissions make special exceptions for municipal buildings, including schools.

If there is no onsite or offsite renewable energy in the ZEC, then the net-zero energy deficits in the ZEC may be offset by adding renewable energy credits (RECs) to bridge the gap between renewable energy sources and uses. The purchase of renewable energy credits that offset carbon use with complementary carbon sequestration—such as planting trees or supporting renewable energy generation—although not ideal, is still considered a worthy goal by NREL.

Reaching that milestone of a zero energy community does not mean that the ZEC must self-sustain net-zero energy. ZECs are about best effort in pursuit of zero-net, not absolute zero.

Future policies that will likely affect the ZEC may include the institution of cap and trade, and carbon tax, which monetize carbon, increasing its ultimate cost to discourage the use of fossil fuel. Expert sources argue which would be most effective and whether either policy should be pursued (Lester 2013). My own forecast cannot provide me any clarity on when such carbon policies may be enacted, but either would be good for ZECs.

Increasingly, regulation falls to the individual company rather than to federal or state guidelines. This is seen in the implications of carbon accounting.

Carbon accounting is the process by which a company uses software programs and on-the-ground monitoring to account for the six major greenhouse gases (carbon dioxide, methane, and nitrous oxide, HFCs, PFCs and sulfur hexafluoride/SF6), as defined under the Kyoto Protocol.

> *Today, carbon accounting is voluntary. Walmart and Coca Cola are among the many businesses that have established their own strategic-level carbon reduction targets. Microsoft and Google each use their own carbon accounting software to measure their corporate carbon footprints. (Everblue 2011)*

### How Does Carbon Accounting Work?

> *If a company builds a giant new warehouse, sends its employees jetting across the country on airplanes and adds new vehicles to their fleet, they are emitting carbon. The "built" environment and transportation are the two largest culprits in corporate carbon emissions. If a company uses environmentally preferable products when building the new facility (or avoids building it altogether), limits air travel and switches to hybrid or electric vehicles, then it reduces carbon emissions. (Everblue 2011)*

I have not yet discovered any state or federal policy that directly addresses ZEC formation. The relatively new concept of ZECs will undoubtedly attract regulation as it develops and spreads. The first steps in such moves toward regulation and ZEC product conformity will likely be through certification, not law (See Section 16: Certification of a ZEC). Policy can greatly affect the rate of adoption of new ZECs. Communities and utilities, or electric co-ops seeking to develop ZECs can do so more easily and beneficially if government regulates energy programs that are complementary to ZECs. Key areas of regulation may come in the forms of new and emerging laws, tariffs, rate structures, district formation regulations, interconnect rules, jurisdictions, financings, bond funds, and other subjects, including: net-metering, solar panel installations, views and shading, wheeling-of-power, tax credits, and rebates.

Even the geopolitical agenda drives the federal government to improve the power grid vulnerabilities through ZECs and similar developments—in order to increase Homeland Security.

One of the most pressing regulatory questions to consider in ZEC development is whether a ZEC is even legal at a particular location or in a particular configuration. Every state has its own public utility commission and unique laws, regulations and methods of enforcement. ZECs per se are likely not addressed specifically by any laws; the generalities of their commercial and technical constructs may, however, be addressed by existing laws.

> *According to Bernays T. (Buz) Barclay, investment banker for power, renewable energy, infrastructure and legal counsel to entrepreneurs and project developers, regulations in twenty-nine states provide the means for in-the-fence power generation. (Barclay 2013)*

In this respect, the border of the ZEC is the "fence" in which the distribution of energy occurs. This is of greatest concern for an off-grid ZEC that must generate all power on-site. Such a community has to plan a legal means for the distribution and sale of electricity within its borders. Backup generators at an on-grid ZEC may also be a regulatory concern in some states.

Because ZECs are in a nascent stage of development today and because regulations affecting renewable energy are also changing rapidly, it is likely that there is little or no available knowledge of the laws pertaining to ZECs available from anyone but local merchant power-industry law experts, who understand the details of the electricity laws related to a ZEC project.

The Department of Energy (DOE) provides numerous Smart Grid, energy efficiency (EE) and renewable energy (RE) programs that provide scientific research, applied research, market intelligence and technology deployment, as well as assisting commercialization of new energy opportunities. DOE provides significant resources through large grant programs and financial awards intended to stimulate technology transfer, entrepreneurial activity and groundbreaking projects. Many of the country's Smart Grid and grid security programs, as well as the National Renewable Energy Laboratory, are funded by DOE.

Regulations can provide initiatives to foster ZEC adoption in all corners of the marketplace, encouraging and discouraging various behaviors of individual consumers, utility companies, communities and governments. Enactment of policy that encourages ZEC developments would provide economic incentives and other social and environmental benefits to the public.

Those interested in joining a ZEC or wanting to finance energy-efficient or renewable energy renovations may benefit from pending legislation regarding mortgage qualifications, bills such as the SAVE (Sustainable Accounting to Value Energy) Act, Senate Bill 1737 (2011), introduced by Sen. Michael Bennet (D-CO) and Johnny Isakson (R-GA) in 2011.

> *As of 2012, during the mortgage approval process, the value of a home's principal, interest, tax and insurance (PITI) is typically calculated on a monthly basis and compared against a borrower's monthly gross income. Mortgage lenders generally prefer a home's PITI to be equal to, or less than 28% of a potential borrower's gross monthly income. (Investopedia 2012)*

*The SAVE act would add utility bills to the PITI equation—making it PITIU—and would allow mortgage financiers to consider an acceptable PITIU score as a higher percentage of the homebuyer's or owner's gross monthly income. Higher PITIU will help homebuyers or owners secure lower interest-rate mortgage financing for homes that are more energy-efficient, incorporate renewable energy or help finance energy-efficient remodeling. (Civic Impulse, LLC 2013)*

However beneficial a ZEC may be to its users, it is also important that policy consideration be given to whether a ZEC unduly burdens a utility, which might create a liability for uninvolved ratepayers. A ZEC renovation of a large neighborhood could cause the utility company's existing substation and distribution equipment to be used less, and therefore, be unable to return revenues on the utility's original investment. In other cases, utility company capital equipment could become completely stranded assets.

Economic development initiatives can improve a region's sustainability, generate sales tax revenue and fortify property tax roles for some entities. For example, in Denver, Colorado, eight hundred new homes are being constructed in the Lowry ZEC, a base-reallocation project that is being built upon land transferred from the US Air Force to the City and County of Denver. The will of the mayor of Denver and his Office of Economic Development are the driving forces behind establishing this project as a ZEC.

This ZEC guide does not suggest how to change regulatory policy, but rather asks those considering legislation or those contemplating political strategy, to thoroughly explore the opportunities of ZECs. I hope that such individuals or groups will utilize their leadership to become involved, educated and actionable, striving to bring the ethical best out of all the associated utilities, communities, and regulatory bodies—in support of ZECs.

When it is time to discuss ZEC policy with politicians, regulators, utilities and community members, I would stress the need for diplomacy. It should be expected that strong and often opposing opinions will be voiced about the practical, political, and financial concerns. Many of us have strong opinions about what the government should, and should not, spend money on, about what should be done by a federal, state or local realm of government, or if government should even be involved.

*What specific policies concerning ZECs are problematic? What are people's opinions about energy security, renewable energy technology and climate change? In addition, what are the sensitivities toward privacy, and how do they vary? These questions and values resonate differently for each person and institution, and change regularly with the political tide, and with climate-related disasters. (Leiserowitz, et al. 2011)*

The goal of ZECs is to give the idea a shot and create an open dialogue, so it is important to create an open framework for that dialogue, a conversation that will meet every group where they are. Skip Spensley, Professor of Sustainability, University of Denver, comments that:

*Understanding and respecting how other people may feel about policy issues that affect ZECs is important. The people affected by a ZEC need to engage in discussion, develop a shared understanding, and explore, and find a middle ground, and have a productive dialogue about the possibilities. They need to ask others about what they know about ZECs and what they think about the practicalities and what types of policy matters would relate to a ZEC. Ask affected people to put their opinions aside and give a chance to ZEC, and explore the practical, financial, technical and ethical considerations of a ZEC. An open mind about the NEEDS among a cross-section of people leads to a common understanding and vision. When people aspire to move forward based on their personal interests, and not just the reasons associated with their role, the inertia of any new project builds. (Spensley 2013)*

Despite varying beliefs and motives, the benefit ZECs bring to a community, beyond sustainability and including energy security, should be enough to warrant exploration. Once a critical mass of interest is developed, the approach can help to bring several sustainability goals into a single, and easier-to-manage, program. After all, ZECs harness energy efficiency, renewable energy, electric vehicle charging infrastructure—and very importantly, behavior modification of energy users.

I recommend that ZEC teams gain familiarity with state and federal legislation and energy policy matters, including grants, bonds, tax credits, and other tax-related provisions, and voice their community's recommendations to elected and appointed officials. Also, ZEC planners should familiarize themselves with programs and subsidies for project developers, ratepayers and energy system manufacturers. If a ZEC is planned to encompass an entire municipality, determine when the next periodic renegotiation of the utility company franchise will occur.

In some jurisdictions, like the City and County of San Francisco, there are programs in place that require energy audits to be furnished by commercial real estate holders whenever property is transferred. Non-compliance can result in fines and penalties.

The Federal Energy Reliability Corporation (FERC) provides regulations that supersede local ordinances when renewable power is being routed to federal facilities, allowing wheeling of power, and uses of the utility infrastructure and rights-of-way that would be considered trespass if executed by private companies.

Some new regulations allow ZECs enlarged options for renewable energy. For example, new regulations in Colorado permit geo-thermal developers to act as utility companies; another provides for community solar farms to be developed.

The regulatory sphere adds an additional layer of complexity and uncertainty to ZEC planning. In consideration of changing policies, ZEC project plans may contain distinct planning alternatives, such as phased development processes, that move in parallel with policy environments.

*Figure 10 - EXAMPLES OF TAX AND INVESTMENT CREDITS, GRANT FUNDS, AND REBATES THAT MAY REDUCE THE COST OF IMPLEMENTING A ZEC PROJECT*

### Renewable Energy Regulations

The NREL ZEC classification system provides for utilizing both on-site and off-site renewable energy supplies. If the electricity regulations in your region provide a ZEC with open-access, that ZEC may simply purchase renewable energy from a distant wind farm, hydroelectric plant, solar installation, or other source of renewable energy—and transmit that energy through another operator's transmission circuits to the ZEC. The Federal Energy Regulatory Commission (FERC) regulates such situations under the Power Grid Access Rule that defines the rules of engagement for open-access to a power transmission grid.

> *Grid Access Rule (1996, as revised in 2007 by Order # 890) was mandated in order to bolster competition in the wholesale power market. (James 2011)*

In order for a ZEC to buy energy through the grid, the ZEC must be entitled to purchase wholesale power from the grid, and the regulations of the federal, state, municipality or other jurisdiction govern that right.

A ZEC can perhaps find dispatchable loads, uses of the energy that occur in order to take advantage of a temporary windfall of renewable energy. There may be means to store energy, or sell it to another user. It is worthwhile to consider local uses for excess energy in order to provide other value.

### Net Metering

Electricity produced from renewable energy such as solar (PV), wind energy, or fuel-cell power is regularly used in conjunction with net-metering policies in effect at that location. Net metering provides a mechanism for sharing electricity cost allocation between a single retail customer and a local electrical utility. Net metering provides for the electricity produced by the retail customer's on-site renewable energy to be used to meet the individual electricity demand of the customer's property, and then to outflow unused electricity to the electrical utility grid.

Net metering is available in most states, and each state's utility tariffs uniquely define what the utility charges and credits the net metering customers. Although the calculation is tied to the net-balance of inflow and outflow of electricity recorded on the electric meter at the location, the inflow and outflow rate can vary and may include limits on the amount of payment a customer can derive from electricity flowing to the grid.

### Virtual Net Metering

Electricity produced from renewable energy is increasingly being used in conjunction with virtual net-metering policies. Virtual net-metering, where available, enables sharing electricity cost allocation between multiple retail customers and a local electrical utility, based on data from many individual customers' electrical meter readings and meters associated with shared renewable energy facilities such as solar gardens.

Virtual net metering provides a means for developing large on-site or off-site renewable energy systems that generate shared electricity for the community members. Like net metering, the net of the inflow and outflow affect the utility company's charges to the individual retail electricity customers. Because data can be used to develop net-metering cost adjustments, an actual connection of electrical circuits to a common electrical meter is not required. Instead, formulas are used to allocate each customer's share of ownership in the renewable energy infrastructure, or to ratify their share of the savings based on an aggregate of meter readings so the credits are reflected on the energy bills of the individual community members.

ZECs are typically large and long-term projects that can provide beneficial social and economic impacts, the sort that cannot escape notice by citizens and officials, and that breed strength from shared goals and the common will of the community. Building and harnessing community and political will and exploring solutions and projects that can positively affect national security, community health, the environment and the economy, provide a result that no politician or party or person should oppose. Fran Treplitz, Energy Program Director, Green America, advocates ZEC development:

> *Zero energy communities is a great term that brings forth the idea that a community should take many actions to come as close as it can towards reaching zero . Green America is working on creating the Clean Energy Victory Bond in support of clean energy and energy efficiency measures. This new, proposed US Treasury Bond, like a Savings Bond, would be available starting at $25, allowing most Americans to invest in our clean energy economy. The funding from the bond will help to extend tax credits for tried and true renewable energy and energy efficiency projects. This incentive is helpful to promoting the shift to renewables. Oil and gas are already heavily subsidized, so this is a move in the right direction, to take a step toward leveling the playing field for renewable energy. (Teplitz 2013)*

*Figure 11 - THE PROPOSED CLEAN ENERGY VICTORY BONDS ACT OF 2013 WILL PROVIDE FUNDS FOR ZECS AND OTHER CLEAN ENERGY PROJECTS. (Courtesy of Green America)*

Renewable energy, available from the sun, wind, tides, biologic materials, and through geothermal heat, replaces conventional energy obtained from nuclear plants, coal, gas, and oil-burning generation plants. Renewable energy is appropriate for providing electricity and heat to buildings and for powering electric vehicles used for transportation.

In this section, I will provide an in-depth explanation of the renewable energy technologies most applicable to ZECs. These specialized technologies comprise a fast-changing sector, with continuous technological improvements and market advancement—so rapid that changes will occur in this sector before you can completely finish this guide. In truth, each of these technologies could fill a book on their own. Nonetheless, as a result of reading this guide, ZEC planners will be able to determine where to focus their research, utilize the extensive glossary and resources, and better prepare for each phase of planning with these technologies, from conception to execution.

As power plants age and need to be replaced, there is a strong potential for renewable energy to displace nuclear and fossil fuels, provide a sustaining economic and environmental benefit and reduce America's susceptibility to natural, human-made, and/or geo-political disasters. Local renewable energy is America's own abundant energy that, unlike fossil fuel, will last forever.

> *Have we reached a tipping-point? Can we afford to move from fossil-fuel to renewable energy supplies? According to Amory Lovins, we have already reached that tipping point. Moreover, his thesis is that it is more profitable for business to use renewable energy than carbon-based fuels. If we utilize renewable energy and use energy more efficiently, the considerable savings will bolster the nation's economy and reduce dependency on foreign oil. Lovins forecasts such a transition with already proven technologies and regulations, without increased subsidies and with rates-of-return and investment that are currently acceptable to businesses. (Lovins, Chief Scientist and Founder, Rocky Mountain Institute 2013)*

> *President Barack Obama concurs with the experts and has consistently called for a unified strategy to diversify America's energy, the ultimate goal of which is to shift to domestic energy. He states that America needs an all-out, all-of-the-above strategy that develops every available source of American energy. In 2012 President Obama said, "It is imperative that America reduce its reliance on petroleum fuel, most often derived from unstable regions of the world." (Obama, State of the Union Address 2012)*

> *Energy was also a primary topic in the 2013 State of the Union address. Energy was mentioned twenty times in relation to national priorities, job creation, national investment opportunity, ability to control our future, automotive efficiency, energy cost, sustainability, energy independence, and the aging power grid infrastructure. (Obama, State of the Union Address 2013)*

> *The US Army holds the same position and worries that supplies of oil will become curtailed either by natural depletion, geo-political sanction, war, natural disaster or man-made accident or attack. (Yergin 2011)*
>
> *As mentioned previously, Katherine Hammack, Assistant Secretary of the Army (Installations, Energy & Environment)), has stated: The Army is quickly mobilizing renewable energy technology because we are concerned with the risks of fossil fuel dependency and need to avoid the energy costs that have already exceeded one-billion dollars during 2011. (Hammack 2010)*
>
> *Renewable energy advocate and author David D. Chiras agrees with this strategy:*
>
> *Rising oil and natural gas prices along with climate change are additional reasons to switch to renewable energy.... Another big advantage of renewable energy technologies is that with a few exceptions, the fuel is free. It is not under the control of oil cartels or wealthy multi-national companies. (Chiras 2006)*

But how will this power generation be achieved? Should we build newer and somewhat improved coal and nuclear power plants, as is being done in the rest of the world today, or shall we build more renewable energy supplies? The World Nuclear Association claimed that:

> *Following a 30-year period in which few new reactors were built, it is expected that 4-6 new units may come on-line by 2020, the first of those resulting from 16 license applications made since mid-2007 to build 24 new nuclear reactors. (World Nuclear Association 2013)*

Conceivably, the same resources invested in ZEC development, instead of nuclear, would offset the need for more nuclear plants, and also allow a continuum of coal and nuclear plant retirements. Sentiments about nuclear energy may change quickly in the aftermath of the leakage of radiation following the nuclear disaster at Fukushima, Japan.

> *Experts are not clear on how the radiation will disperse. The appearance of radiation offshore from Japan would likely spark global protest and debate regarding further utilization of nuclear energy. (Jones 2013)*

The US power grid and its associated power plants are the second largest consumer of fossil fuels after transportation. Today's electric grid is primarily fueled by coal, and the power plants contribute greenhouse gases and heavy metals to our environment. The bi-products of these operations include carcinogens, tetrogens and mutagens that threaten the environment and human health.

The way we use energy today could not have been imagined in the late nineteenth century, the time that my great-grandfather, James Royal Ingoldsby, lived. Not imagining the population explosion in tandem with depleted resources, we adopted fossil fuels (then superior to whale oil) and committed to following the centralized power generation design established by Thomas Edison. Changing the models of electricity technology and markets today requires nothing short of letting go of these traditional models of producing and distributing energy.

ZECs provide a tremendous opportunity for utility companies to test new renewable energy and control technologies, and to develop business models that engage utility and retail customers in new models of rate structures, expanded service offerings, and customization according to consumer preferences. There are a number of options to fossil fuels that have emerged as solutions to our hunger for oil.

*Nuclear energy does not seem like an option in the wake of several highly publicized nuclear accidents. The American public better understands the consequences of nuclear accidents now, making nuclear too big a risk for America unless there is a breakthrough in safety. Before we give up hope, I want to cautiously acknowledge that proponents of Thorium Molten Salt technology have recently claimed the technology is safe, affordable, and promises environmental benefit. (Horsting 2013)*

The prospects for coal-burning generation are severely diminished due to new regulations of the Environmental Protection Agency (EPA) during 2012. As a result, the futures prices for coal have risen multifold since 2012, and numerous coal burning power plants have already been decommissioned or converted to natural gas operation. Building new coal plants, or so-called "clean" coal plants, is wrought with resistance. Although increasing prices for carbon fuels may make renewable energy and energy efficiency more attractive, the changes in regulation may result in significant cost increases for electricity.

*The market-clearing price for new 2015 capacity – almost all natural gas – was $136 per megawatt. That's eight times higher than the price for 2012, which was just $16 per megawatt. In the mid-Atlantic area covering New Jersey, Delaware, Pennsylvania, and DC, the new price is $167 per megawatt. For the northern Ohio territory served by FirstEnergy, the price is a shocking $357 per megawatt. (Kerpin 2012)*

The power utility industry can continue to shift from coal to natural gas, increasingly more plentiful and affordable because of windfalls unearthed through the use of hydraulic fracturing. The use of natural gas is applauded by many as a domestic fuel. Conversely, growing concerns about the environmental risks of hydraulic fracture drilling are placing pressure on that industry.

Businesses already invested in nuclear and carbon-based energy generation that own gas pipelines and power transmission lines, have plants that are not at the end of their viability and may find it hard to operate in a new energy economy.

Renewable and distributed energy generation conspire to make all the power and utility companies' assets become less valuable and potentially irrelevant, or extinct. Change would be easier if the nation's electrical system could be developed from scratch. Today we are bolting solar panels onto our problematic grid, and the ZEC is a step in creating zones where the future can be better understood and enjoyed and lead the way of change for those utility companies that wish that change was not necessary.

With increasing demand to support state renewable portfolio standards (RPS), new electric vehicles (EVs), aging power plants (nuclear, coal, and oil), and sustainability charters by business, institutions and government, I expect the use of renewable energy will be embraced, making it the dominant source of energy in the US. Achieving this could make America the model for other countries to follow.

The growing consumption of oil and gas is the single most polluting of humankind's activities and transportation is the principle user of fossil fuels. Electric vehicles that can derive their power from renewable energy in the form of electricity provide an extremely advantageous opportunity for reducing the use of fossil fuels.

Conservation can play a big role in bridging the gap between the current energy supply and growing requirements for more energy, but the US power supply must grow to support the additional energy that will be required by tens of millions of new electric vehicles that will make a significant appearance in the US during 2013. (Note: See Section 10: Electric Vehicles)

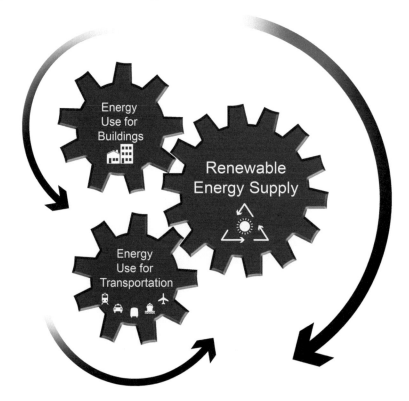

*Figure 12 - A RAPID SHIFT TO RENEWABLE ENERGY INCLUDES ELECTRIC POWERED BUILDINGS AND VEHICLES, SIMULTANEOUS WITH A SHIFT FROM AGING NUCLEAR, COAL, AND OIL GENERATION PLANTS, THEREBY ACHIEVING PROFOUND ENVIRONMENTAL, ECONOMIC AND ENERGY SECURITY BENEFITS.*

### *Renewable Energy Economics*
The primary economic benefits of renewable energy are that it is (1) naturally available, (2) in an abundant supply, (3) available for an endless period of time, and (4) free to use. The other factors that affect the economics of renewable energy most meaningfully are as follows:

(1.) The market for renewable energy technology has grown significantly, dramatically reducing equipment costs, due to benefits-of-scale, better performance, easier deployment, and better educated consumers.

(2.) Soft-costs associated with engineering, permitting, and equipment installation for renewable energy equipment installations are also falling because of increasing familiarity and advocacy by officials and more efficient permiting and inspection processes.

(3.) The financial services industry now provides accessible financing, loans and leases, to amortize the up-front cost of renewable energy equipment installations. This reduces the need for the consumer to outlay capital in order to receive the benefits of renewable energy.

(4.) Financial incentives, including subsidies, tax credits, and rebates, encourage national goals for developing alternative energy, affording purchasers with benefits that improve returns-on-investment in renewable energy.

(5.) Regulation is trending to the benefit of renewable energy, opening grid power access, stimulating competition, restricting pollution, and permitting innovative rate structures and net metering, as well as solar garden development.

(6.) *For all these reasons, renewable energy promises to secure the nation's energy supply, provide environmental restoration, and re-circulate the funds spent for energy within the US economy, providing an economic boost for the nation as we save an estimated five billion dollars per day that is currently spent as a result of dependence on foreign oil. (Lovins, Reinventing Fire: Bold Business Solutions for the New Energy Era 2011)*

(7.) Proven performance: Renewable energy technologies (solar, wind, geothermal, fuel cells, tidal, biogas, hydropower, waste-to-electricity, and naturally occurring methane) have been deployed widely and have demonstrated reliable performance and economic viability.

Where will renewable energy lead the nation today and tomorrow? What technology developments will be breakthroughs? How will business models evolve? The answers to these questions are speculative.

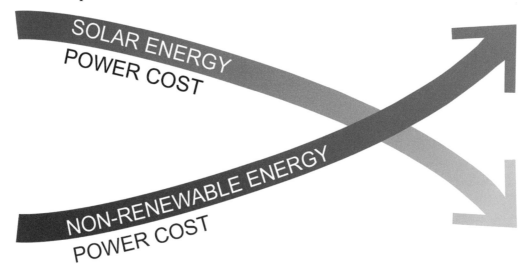

PAST                                                                           FUTURE

*Figure 13 - THE JANUARY 2013 EDISON ELECTRIC INSTITUTE REPORT INDICATING NON-RENEWABLE ENERGY POWER COST CONTINUES TO INCREASE AT THE SAME TIME AS SOLAR ENERGY COST IS RAPIDLY DECREASING. (Kind 2013)*

The cost of conventional energy is rising steadily, and the cost of renewable energy generation is dropping even more rapidly.

### Renewable Energy for ZECs

Many promising technologies become practical to operate at different scales. Changes in market, financial considerations and regulations addressing the use of renewable energy are sweeping the country, even as I write this guide. The component technologies and system configuration of renewable energies are constantly improving by degrees. A future watch is warranted, especially given the fast-changing nature of the new energy economy.

Renewable technology is locational, and the potential for each type of renewable energy technology must be evaluated for the exact site where it is to be deployed in order to derive reliable estimates of the expected energy generation performance at that site. In the case of new development, it is even necessary to consider the shadows that will be created by the "built" environment and how wind vortexes caused by structures may positively or negatively affect wind generation.

Maps are available from most state energy offices that indicate wind, solar, thermal and hydropower potentials around a given state. Other maps, like those available from NREL, depict the country and world. These maps provide only a general idea regarding renewable energy potential.

In the remainder of this section, I have limited my discussion of alternative renewable energy to solar electric, solar thermal, small wind, geothermal, fuel cells, hydroelectric, biofuels, and waste-to-energy, appropriate to a ZEC. These particular renewable energy sources are the most prevalent renewable energy technologies today.

In Section 7: Understanding the Electric Grid, I introduced the concept of intermittent energy and how that affects a regional power grid. However, not all renewable energy is intermittent. Solar and wind renewable energies operate intermittently, but geothermal, fuel cells, hydroelectric, biofuel, and waste-to-energy forms of renewable energy are usually not intermittent in operation. Renewable energy from solar panels and wind generators produce intermittent energy. The number of hours of sunshine and clouds affect solar generation. Storms and windless periods can reduce wind generation, and wind is most prevalent at night.

Renewable energy from geothermal, hydroelectric, fuel cells, biofuel, and waste-to-energy energy sources can operate continuously. However, these sources would subside if there were an equipment failure, limited water supply, or fuel supply problems.

*Figure 14 - LOPEZ COMMUNITY LAND TRUST'S NET-ZERO ENERGY AFFORDABLE HOUSING NEIGHBORHOOD ON LOPEZ ISLAND, WA, DESIGNED BY MITHUN  (Photo courtesy of Mithun, photograph by Juan Hernandez)*

### Solar Electric - Photovoltaic Solar Technology

Solar photovoltaic (PV) panels provide an affordable and sustained method to convert sunlight into electricity with panels mounted on the ground, on rooftops, or on the topside of parking canopies. They are usually flat, though sometimes configured in troughs with reflectors, steerable collectors, and other form factors. There are abundant choices, such as roof tiles with integrated PV and curved PV panels and glass designs incorporating PV. Photovoltaic solar panels produce direct current (DC) electricity that is converted to alternating current (AC) power through an inverter.

In the northern hemisphere, we look to the south skies for sunlight. The exact position, azimuth and elevation of the sun in the sky changes hourly and seasonally. The intensity of the sunlight varies depending on the latitude where solar panels are placed. As latitude increases, moving north from the equator, sunlight projects through increasing distances of the earth's atmosphere, lessening the energy potential in northern latitudes.

Solar energy is produced intermittently; however, solar energy supply and community energy demand generally correlate favorably, as solar is typically most productive during sunny periods, the same times of day as the energy demand for refrigeration and air conditioning are also at their peak.

Communities and individuals who use solar electric energy should do so as an investment, expecting both a financial return and environmental restoration benefits as results. Commodification of solar photovoltaic equipment markets during the early part of this century caused equipment costs to become a fraction of prior costs. Solar photovoltaic is now a low-cost renewable energy, suitable to most US climate zones. The break-even for solar is calculated in consideration of the original installed equipment cost, the amount of sun available at the location, the prevailing cost of electricity in that locale, and available tax credits, rebates, or other incentives. The break-even period is therefore shorter in sunny states with high electric rates and more generous incentives than other locations.

### Solar Thermal Energy Systems

> *In the US, more than twenty-five percent of the energy used in buildings and more than fifty percent of the energy used in residences, is for heat. (US Department of Energy, Office of Building Technology - State and Community Programs 2008)*

Solar thermal is therefore an appropriate source of energy to provide space heating and swimming-pool and hot-water heating for residential and commercial buildings.

Solar thermal energy systems produce heat instead of electricity. The basic system incorporates solar collectors, piping, storage/transfer tanks, and control systems. Solar thermal panels, like solar photovoltaic (PV), require orientation to the sun and the presence of sunshine.

The systems vary in size, for use in small buildings and even for community-scale projects that deploy high-temperature designs that are either dish-shaped, linear field or power-tower types. Systems are available for use in both southern and northern states, and provisions to prevent freezing are available.

Most large-scale solar thermal systems, like those developed by the BrightSource Energy Company, are built in extremely large fields, using miles of ground-mounted mirrors that reflect the sunshine to a central tower where heat is captured by circulating fluids that then heat water to create steam to power turbines—which in turn, provide electricity for large regions and large cities. (Too big for a ZEC?)

Solar thermal is used in conjunction with insulated hot water tanks capable of storing heat that is generated in the daytime. These are an efficient storage device and make solar thermal's intermittent operation almost forgivable.

### Swimming Pool Heating

The greatest use of solar thermal energy is to heat swimming pools using low-temperature type collector panels. The application provides the potential to significantly reduce energy cost, extend seasonal use of swimming pools and restore environmental conditions.

### Building Heat and Hot Water

Solar thermal heating and hot water systems used at residences and commercial buildings represent the primary use of the technology today. The energy gathers at the collector panels and is circulated (using solar powered pumps) through pressurized and non-pressurized pipes filled with water or glycol effluents to heat storage and transfer tanks. Other pipe-loops are set up to connect the medium-temperature solar thermal collectors with heating-ventilating-air-conditioning (HVAC) equipment, to heat domestic hot water, and to transfer heat to radiant-heated floors. Radiant floor heating systems use special piping that snakes through a concrete floor slab. The concrete floor radiates heat into the living space, and the concrete also stores heat for when the sun goes down.

Solar thermal-heated water flows into a new or existing hot water heater through the water supply connection. If there is inadequate solar activity, the traditional hot water heater activates until the water is again hot.

### Solar Thermal of Yesterday and Today

Solar thermal technology gained a poor reputation for its technical performance when it was first introduced to US markets in the 1960s through the 1980s. Many of the older systems were reported to be prone to leakage and deterioration.

The technology has improved dramatically compared to the experimental pioneering installations of the previous century. Modern control equipment now adds to the efficiency and safety of these thermal systems, and the control systems provide reports regarding the amount of energy generated, and the exact savings in electricity. These new control systems coupled with advances in material sciences that prevent leaking and deterioration of the panels have satisfactorily remedied the problems of the past, and can even allow an owner's thermal system to automatically call for maintenance service if maintenance is required.

> *According to a confidential business source who is of national prominence in the industry, thermal storage has been found to be considerably less expensive than electrical heating in many major US cities, and many of these cities are undergoing retrofits to solar thermal as part of utility-wide rollouts. (Undisclosed 2013)*

The cost of solar thermal equipment systems is often much lower than other types of renewable energy system deployments. When installed in regions where electrical power rates are high and solar energy is available (most US climate zones), solar thermal systems provide an accelerated cost payback.

### Wind Energy - Small and Large Wind Turbines

Wind turbines powered by wind energy produce mechanical energy that can drive electrical power generators or operate pumping or grinding equipment. Wind turbines produce intermittent energy, in that they only produce electricity when the wind is blowing. Storms and windless periods can reduce wind generation. Wind is most prevalent at night.

The three primary types of wind turbines are Savonius and Darrieus vertical axis wind turbines (VAWT) and horizontal axis-towered wind turbines (HAWT).

The performance of wind turbines is expressed as average energy output (AEO) at a 10 mile-per-hour wind speed.

### Small Wind

> *Small vertical wind turbines of 7 - 25 feet are deployed for remote power requirements and distributed energy generation (DEG) applications, such as at a ZEC. Small wind turbines supply 300 -10,000,000 watts of electricity. The spacing intervals used with small vertical wind towers is typically a distance that is six to ten times longer than the turbine rotor diameter. (Meneveau and Meyers 2011)*

A widening choice of options is being revealed as new products prepare to enter the market, including turbines with higher efficiencies, solar photovoltaic integrated in blades, and vertical axis turbines suited to streetscape installation in dense urban environments.

*Industry analysts Navigant Research and Pike Research have projected a sharp and sustained increase in the small wind marketplace from 2009-2015, providing an overall increase of two hundred sixty three (+263%) cumulative annual growth of revenue. (CAGR) (Asmus 2012)*

Is small wind an option for your ZEC? ZEC planners should consider whether wind energy is available, and if so, where there space is available to use towers, mounts integrated with buildings, and/or street-scale wind turbines integrated with greenbelts, parks, transit corridors and/or pedestrian malls within communities.

### Big Wind

Big wind should be considered as an energy source for ZECs located in wind-rich locations, and when space is available to develop wind towers or wind farms, big wind could provide a significant source of energy. An island ZEC supported by wind turbines on the water, or a large municipality becoming a ZEC and installing wind within transit corridors or greenbelts, are examples of such application.

Wind power has been heavily commercialized in the big wind arena, with billions of dollars-worth of equipment installed in the US already. Big wind might be perfect except for its size, its threat to birds, and its tendencies toward intermittent operation. The intermittent nature of wind power is perhaps the technology's most significant drawback.

*Big wind turbines range in size from 150 - 650 feet. They are usually deployed on wind farms on land and ocean, for grid-scale power generation. The spacing intervals used with big wind towers is typically a distance that is fifteen times longer than the turbine rotor diameter. (Meneveau and Meyers 2011)*

The environmental concerns related to wind may present challenges to ZEC development. The roadways that allow the site delivery and installation of a big wind tower sometimes require a wide clear-cut road, possibly through woods that are in the pathway. The impacts of such clear-cuts are considered significant by environmentalists.

*Another concern of big wind installation is the danger presented to birds and other wildlife. The City of Boulder's Department of Parks and Wildlife is currently making such an assessment and is determining how the risks can be managed through integrating consideration of habitat, plus flight and other paths used by wildlife. In all development, attention to the subtle patterns of nature, including light, shadow, wind, water, wildlife movement patterns, and natural riparian structures, can soften such impacts, and lead to legacy place-making. (Design Workshop 2007)*

The costs of equipment and installation required for small and big wind is continuously diminishing. Where permissible by law, a ZEC may choose to purchase wind power from a distant wind farm. The range of products, sizes and form factors is improving. The performance of wind turbines is increasing, permitting productive use with slower wind speeds—all of this while the cost of conventional energy sources is continuously rising.

A comprehensive evaluation of wind potential, site conditions, environmental impacts, selection of advancing technologies, and cost comparison with prevailing power rates will assist in determining whether wind is applicable to a ZEC location.

### Geothermal Energy

There are several types of geothermal energy systems which range in scale from a single residence to monumental power plants. Among the largest is the Geysers Geothermal Plant owned and operated by Calpine Energy Corporation (Calpine Corporation 2012) that provides 23% of all the renewable energy used in the state of California.

The energy recovery systems utilize either the thermal mass of the earth, or hydrology systems such as deep-well geothermal to drive steam-turbine electrical generation plants; both of these methods provide a means to distribute electrical or thermal power to a ZEC.

When I began my research, I found the case study literature published by the GeoExchange, an industry group, to be very helpful. Later, well-drillers and system installers took me to inspect residences and buildings where geothermal heat pumps had been installed. I was able to speak with families who used geothermal as their source of heat, air conditioning, and hot water. These people were clear that the technology worked for them.

Geothermal energy systems cover a broad range of applications. Whether it is a short loop of plastic buried to exchange energy with a heat pump located at a single family residential air conditioner, a district system providing energy to a community, or a pair of wells that are 1 - 7 miles deep (and powering large electrical power plants that in turn power large regional grids and cities), the applicability for geothermal energy is significant when it is available and developable.

### Deep Well Geothermal Energy

During May of 2012, I visited NREL and saw a demonstration using a large scale model of a deep-well system. Two deep wells are required: a 1 to 5 mile-deep "resource well" to extract hot water and steam and another 5 to 7 mile-deep "injection well" to replenish the water supply released by the resource well.

> *This type of energy plant is large and consumes significant quantities of water. The Calpine geothermal plant at the Geysers in Northern California utilizes wastewater piped from surrounding communities for injection water. Calpine claims that the waste water is no longer threatening streams and rivers and that the process purifies the water and returns it to the atmosphere. (Calpine Corporation 2012)*

The main drawbacks to deep-well geothermal energy have been instances in which chemicals and/or toxins contained in the underground water emit pollution into the atmosphere. Also, salts and other chemicals contained in the geothermal well water can corrode equipment. Like the hot springs that delightfully, but rarely occur, geothermal energy is not available in most locations. The discovery cost is high because of the drilling cost for two deep wells. Concentrated in western states, the energy is currently generating over three million watts of power.

### Geothermal Heat Pumps

In areas where the ground is drillable on an affordable basis, not solid rock, the feasibility for geothermal heat pumps exists. Mike Ryan, President of PanTera Energy, L.L.C, a geothermal developer and utility company, stated:

> *A few feet beneath the surface, the earth's temperature remains fairly constant year round; along the Colorado Front Range that temperature is between 52 - 55 degrees. Geothermal heat pumps, also called Geo-Exchange systems, take*

*advantage of this constant temperature to provide extremely efficient heating
and cooling. In winter, a fluid circulating through pipes buried in the ground
absorbs heat from the earth and carries it into the home or other structure. The
geothermal heat pump system inside the home uses a heat pump to concentrate
the earth's thermal energy and then to transfer it to the interior space for warmth.
In the summer, the process is reversed: heat is extracted from the air in the house
and transferred through the heat pump to the ground loop piping. The fluid in
the ground loop then carries the heat back to the earth. The only external energy
needed for a geothermal heat pump is the small amount of electricity needed to
operate the heat pump, ground loop pump and distribution fan or pump for a
hydronic system. (Ryan 2012)*

Geo-Exchange produced case studies for homes, housing developments, apartment complexes, offices, warehouses, and retail buildings, including a Department of Energy presentation at Fort Polk army base, the largest geothermal installation in the world. These case studies can be located in the appendices.

## *Energy System Components*

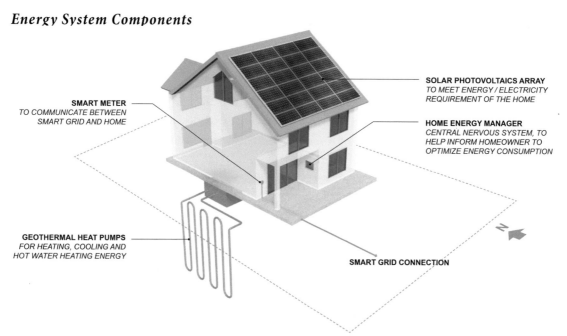

SOLAR PHOTOVOLTAICS ARRAY
*TO MEET ENERGY / ELECTRICITY
REQUIREMENT OF THE HOME*

SMART METER
*TO COMMUNICATE BETWEEN
SMART GRID AND HOME*

HOME ENERGY MANAGER
*CENTRAL NERVOUS SYSTEM, TO
HELP INFORM HOMEOWNER TO
OPTIMIZE ENERGY CONSUMPTION*

GEOTHERMAL HEAT PUMPS
*FOR HEATING, COOLING AND
HOT WATER HEATING ENERGY*

SMART GRID CONNECTION

*Figure 15 - DIAGRAM OF A SINGLE-FAMILY RESIDENCE FEATURING SOLAR PANELS,
HOME ENERGY MANAGEMENT CONTROLS, A SMART METER, AND GEO-THERMAL
ENERGY UTILIZING GROUND-SOURCE HEAT PUMPS WITH AN INDIVIDUAL WELL.
(Illustration courtesy of Design Workshop, Inc. and Lowry Redevelopment Authority)*

The diagram above illustrates the geothermal heat pump's pipe loops that are filled with a heat transfer medium connected to heat pumps and heat exchangers located in the building. This system provides heating, cooling and hot water—without the need for natural gas or additional air conditioning. The piping is filled with antifreeze, which is circulated through the piping to prevent the heat exchange medium from freezing.

In Denver, Colorado, a 300-foot deep vertical well provides for one-and-a-half tons (1-1/2 tons) of thermal energy. A large single-family residence, or a large building, requires multiple wells. Plastic piping of the same type used for underground natural gas pipe is laid into the wells, and the wells are packed with a natural aggregate that provides efficient thermal conduction between the pipe and the earth.

It was a struggle for me to understand how the ground-loop systems can provide both building space heating and cooling, and simultaneously provide energy for continuous hot water heating. I learned that the heat pump can compress and release the fluid and produce heat or cooling and that the equipment always heats the domestic hot water system.

For new buildings and residences, geothermal heat pump systems can approach parity with the cost of other common HVAC and hot water heating systems, assuming reasonable ease of well drilling in the region where it is installed. For renovation however, the costs are greater because the existing HVAC equipment must be entirely replaced.

GeoExchange's website reports their ongoing attempts to convince the US House of Representatives and the Ways and Means Committee to extend until 2020 the current tax credit that expires in 2016. They cite the case for this below:

> *Yet even though geothermal heat pumps (GHPs) could achieve vast energy, economic and environmental benefits across America if installed for all suitable buildings, the technology is still relatively nascent and has been slow to catch a foothold in the broader HVAC market. At current rates of installation (<100,000 average 3-ton capacity residential units per year), GHPs represent less than 2% of the total HVAC marketplace. Reason? Higher 'first cost' incurred by drilling or excavation to place its ground-source heat exchange loop system near the building(s) which a GHP system serves. (GeoExchange 2013)*

The above suggests that up-front costs are an impediment to mass-adoption of geothermal heat pumps. This is debatable, as I see the costs as close to parity for new construction in my home state of Colorado, and yet there is little market adoption. The general public seems to be unfamiliar with this type of energy, and tentative as a result. Developers of the Lowry ZEC expressed concerns about the potential for environmental problems from geothermal wells; however, no such problem at another site had been reported. I followed up, asking an engineer and a well driller about this issue. This was during a research field trip to inspect the geothermal heat pump installation at a large home north of Denver, Colorado. I learned that the geothermal heat pump wells are classified as a "Class-A" well.

The Class-A well is drilled and cased with concrete to prevent seepage of liquids and dirt within the well shaft, outside the well casing pipe. After the hole is bored and the casing pipe is installed, concrete is pumped down the casing pipe until it rises around the outside of the casing pipe and fills the area between the outside wall of the casing and the drill hole for the full depth of the well. Once the concrete is set, the well is redrilled inside the casing pipe. When completed, ground water and contaminants cannot flow down through any part of the drill hole to the subsurface area. Water wells require permitting and inspection through the agencies having jurisdiction in that region

Vertical wells are not the only format of ground source geothermal. Lateral pipes can be laid below building slabs, and loop-fields can be encased in concrete foundation piers and beams. A combination of vertical wells and lateral piping can also be used.

*In the town of Aspen, Colorado, where geothermal is a common source of energy for many of the residences, there is a common practice of laying additional piping loops beneath paved surfaces of driveways and walkways. When those loops are activated, hot water is circulated in order to melt snow and ice away from the driveways and walkways. (Monteux Energy 2012)*

The feasibility of geothermal is a regional consideration, but in those areas where geothermal resources are proven, they can greatly reduce energy needs. Geothermal systems can be installed for each building in a ZEC or as a district system shared among multiple buildings in the community.

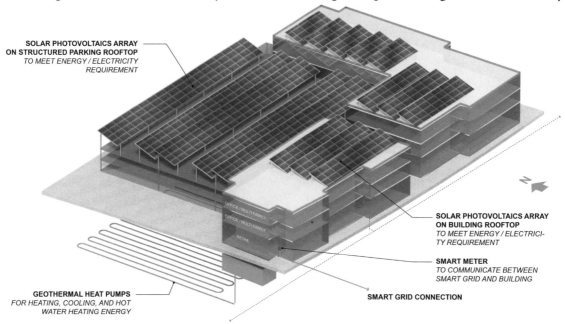

**SOLAR PHOTOVOLTAICS ARRAY ON STRUCTURED PARKING ROOFTOP** *TO MEET ENERGY / ELECTRICITY REQUIREMENT*

**SOLAR PHOTOVOLTAICS ARRAY ON BUILDING ROOFTOP** *TO MEET ENERGY / ELECTRICITY REQUIREMENT*

**SMART METER** *TO COMMUNICATE BETWEEN SMART GRID AND BUILDING*

**SMART GRID CONNECTION**

**GEOTHERMAL HEAT PUMPS** *FOR HEATING, COOLING, AND HOT WATER HEATING ENERGY*

*Figure 16 - MULTI-USE BUILDING AT THE LOWRY ZEC WILL INCLUDE RETAIL STORES, APARTMENTS, AND OFFICES. EXTENSIVE USE OF SOLAR CANOPIES PROVIDES ENERGY AND THE CONVENIENCE OF COVERED PARKING, HELPING TO KEEP SOLAR EQUIPMENT AWAY FROM THE PARK-LIKE GREEN SPACES. (Illustration courtesy of Design Workshop, Inc. Denver, CO)*

### District Loop Fields

In the case of a community, a district-type geothermal system should be considered. The primary distinction is that a district system utilizes one or more loop fields that are shared with many homes and buildings. The advantages include reduced cost of systems and redundant pipe-loops that can be isolated and managed through manifolds, improving reliability and supporting maintenance activities.

The disadvantages of a district loop-field design are that it cannot be phased in at the same rate as new home construction and occupancy in a Greenfield ZEC, and may therefore, require large expense before the customers are settled in and ready to pay for the energy.

District geothermal is, therefore, the best fit for the renovation of an existing campus or community as illustrated below. [Please see illustration on next page (Ball State University 2013)]

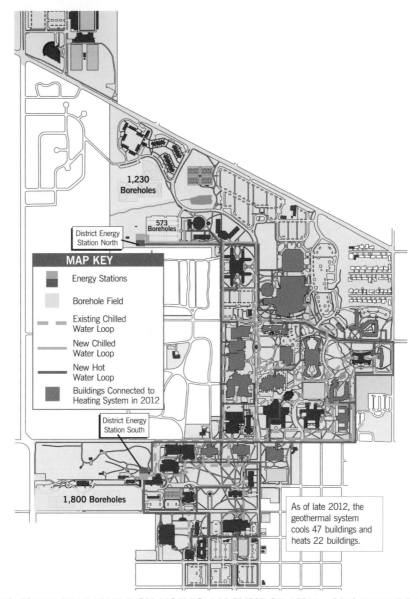

*Figure 17 - GEO-THERMAL ENERGY USING AN INTEGRATED GROUND-SOURCE HEAT PUMP THAT IS OPERATED BY SOLAR ELECTRICITY  (Illustration used subject to the terms of use of Ball State University)*

### Fuel Cells

Fuel cells convert energy developed through chemical and bio-chemical reactions into electricity through an electrochemical process. A fuel cell is an energy transfer device that can operate on renewable or non-renewable fuels.

A fuel cell is much like a battery, one that never discharges, because the fuel cell provides a dependable supply of continuous power. Each fuel cell utilizes an anode and cathode immersed in an electrolyte at the core. Unlike a battery, the fuel cell requires a constant inflow of fuel and oxygen into the chamber of the cell.

*Fuel Cells 2000, a non-profit industry organization states: Researchers continue to improve fuel cell technologies, examining different catalysts and electrolytes in order to improve performance and reduce costs. New fuel cell technologies, such as microbial fuel cells, are also being examined in the lab. The organization's website (www.fuelcells.org) lists thirty-nine fuel cell developers, and indicates several products are available today. (Breakthrough Technologies Institute 1993)*

The number of product offerings is increasing dramatically according to an International Energy Agency (IEA) report by the Committee on Energy Research and Technology (CERT) in November 1998, which listed only one commercially available fuel cell product available worldwide. Fuel cells are combined in serial and parallel electrical circuits in what are called "stacks." Large stacks can be utilized to power buildings, residences, telecom systems, portable equipment and a variety of conveyance and transportation applications (passenger vehicles, buses, trucks, materials-handling equipment). The fuel cells can also be used to produce and store hydrogen.

I first learned about fuel cells in 2000, when I was asked to visit the First National Center in Omaha, Nebraska, to observe fuel cell equipment housed in a large room on the lower floor of their 22-story building. I was amazed to find out that these fuel cells were sufficient to continuously power the entire complex.

*Two years later in 2002, I first grasped the tremendous potential for hydrogen fuel-cell powered automobiles, when presenter Amory Lovins, Chief Scientist and Founder of the Rocky Mountain Institute, explained the subject to the one-hundred leaders gathered at The Economist's Second Annual Innovation Summit and Awards in San Francisco, CA. (Lovins, Innovation in Efficient Vehicle Design and Energy Supply 2002)*

What struck me was that the fuel cell operates constantly; I was amazed to find out that these fuel cells were sufficient to continuously power a residence or building, even while the vehicles are parked.

*Another informative field observation occurred at the Consumer Electronics Show in Las Vegas during 2012. I saw a live demonstration of several fuel cell systems that were integrated in cabinets, each the size of a typical refrigerator. The representative reported that Panasonic had successfully completed more than 5,000 installations of the fuel cells in residences in Japan as part of a government-sponsored program following the devastating Fukushima Daiichi nuclear disaster. (Demonstration 2012)*

*Most large stationary fuel cell systems are fueled by natural gas, but anaerobic digester gas (ADG), derived from wastewater, manufacturing processes, or from crop or animal waste, is being used more frequently as a feedstock. ADG-powered fuel cells are being used at a number of wastewater treatment plants, as well as at breweries and agricultural processing facilities. This up-and-coming resource is counted as a renewable fuel in several states. (DuVivierr 2011)*

Better means of manufacturing, distributing and storing hydrogen and other renewable biofuels will likely make fuel cells even more attractive to ZEC planners. Because the future holds unknown promises, the ability to transition the fuels used to power fuel cells is yet another benefit of the technology.

Environmental concerns always flare when chemicals are utilized in an industrial process. Concerns include preservation of natural feedstock, potential for pollution, and use of embodied energy. Because there are a wide variety of chemistries used in a fuel cell, ZEC planners should evaluate and take precautions to reduce the use of products built with non-recoverable precious metals, non-recyclable or toxic substances.

Fuel cells have undergone enough field testing to verify performance, even in regard to powering sensitive mission-critical data centers. The challenge in the last two decades has been about the economics of fuel cells, the cost per watt. In light of the continually rising cost for electricity in the US, and the lowering cost of fuel cells, the market is beginning to grow. Fuel cells are a technology for ZEC planners to strongly consider.

### Hydroelectric - Energy from Creeks, Rivers, Reservoirs, and Tides

*Hydroelectricity provides the majority of the world's renewable energy by means of harnessing the force of falling water. The US Geological Study (USGS) reported that hydroelectricity represented 7.1% of the nation's renewable energy, of which 5.7% was produced in the US and 1.4% was provided by our neighbor Canada. (US Department of the Interior 2012) (Worldwatch Institute 2012)*

Since the suitable sites for development of additional large hydroelectric dams have been taken already, new development of hydroelectric projects is trending down to a standstill. Instead, the new opportunities are to develop small hydro facilities capable of providing up to ten megawatts in order to serve single communities. To accommodate these community-scale hydroelectric solutions, several types of facilities are available:

(1.) Run-of-the-river hydroelectric renewable energy generation uses small or no reservoir capacity. This technology holds the tremendous potential to provide 13.7% of all of the nation's power requirements, assuming continuous availability of the water resource and the 2011 electricity usage level of the US.

(2.) Tide power plants capture the force of tides, dependably producing electricity from reservoirs—or the less common undershot, waterwheels. Unfortunately, besides the UK where a potential exists to generate 20% of their electricity requirement from tidal power, there are few other sites in the world that have shores suitable for tidal power plant development.

(3.) Underground power stations are built at a point well below an elevated reservoir that is connected with a pipeline. Underground power stations outflow water into a tailrace that is extended to a second waterway or reservoir below.

Proven by more than a century of service quality, hydroelectric has demonstrated its technical and economic viability. Small hydro power is poised to provide a significant contribution of non-intermittent renewable energy to communities located where a combination of water resources and dynamic terrain exist.

Environmental impact analysis (EIA) assessments identify and recommend remediation and offsets for all matters of concern regarding hydroelectric power stations' interaction with natural systems. Particular concerns include: the risk of drought, damage and loss of land, reservoirs filling with silt, changes in natural riparian flows, loss or modification of passage and habitat of wildlife, and emission of methane.

### Biofuels

A biofuel is a type of fuel that is derived from natural materials such as plants, animals, or digested organic waste like manure or matter contained in wastewater. Biofuel is sometimes used independantly, but most fuels today are a combination of a fossil fuel and a biofuel, such as ethanol (a biofuel) additive to gasoline and biofuel combined with diesel fuel to create biodiesel fuel.

> *Biofuel is considered carbon-neutral because the biomass absorbs roughly the same amount of carbon dioxide during its life as when it is burned. Some of the biofuels currently in use include: (Bratley 2007)*

(1.) Biobutanol

(2.) Biodiesel

(3.) Bioethanol

(4.) Biogas

(5.) Vegetable oil

> *To date, most of the focus of the biofuel industry is related to transportation versus stationary application fuel. (Investopedia 1999)*

Some biofuel occurs naturally as a result of offgassing by a body of water or swamp, or the emission of methane from a landfill, or biogas occurring in a cavern or mine. Most biofuels are produced through refining processes that use crop waste (from corn or sugar cane production, for instance), or animal processing waste—or they are themselves a feedstock.

> *Algoil is produced by growing algae in glass tanks filled with water and feeding it carbon-dioxide waste. The algae waste product is used as a fuel, called "algoil." This process is called "carbon sequestration" and also produces oxygen as a by-product. (Sears 2012)*

Biofuel, however promising, has received considerable criticism regarding the following:

(1.) Biofuel often uses food products as feed stock, and that process can inflate food cost.

(2.) Biofuel is less efficient than may be apparent because there may be significant embodied energy incurred during the fuel production.

(3.) Many biofuels (including biodiesel) thicken, becoming more viscous, making it potentially troublesome in cold weather conditions.

I like to believe that these problems will be rectified and that natural gas-fired power plants will convert to carbon neutral biofuels in the future. Even though the opportunity to use biofuels in ZECs is limited today, biofuel should be a defined element in any ZEC planning team's future.

### Biomass

> *Not to be confused with biofuel, biomass is a source of renewable energy that includes organic sources derived directly from wood, crop waste, and other living matter. The vital difference between biomass and fossil fuel is time scale. (Biomass Energy Centre 2013)*

There are several categories of biomass, including:

(1.) Virgin wood: wood from forestry, arboricultural activities or from wood processing

(2.) Energy crops: high-yield crops grown specifically for energy applications

(3.) Agricultural residues: residues from the harvesting or processing of agriculture

(4.) Food waste: waste products from food and drink manufacturing, preparation and processing, and post-consumer waste

(5.) Industrial waste and co-products: products from manufacturing and industrial processes

(6.) Biomass can be converted to a fuel through combustion, gasification, or pyrolysis using combinations of heat and chemicals in the reaction. The most likely uses of biofuel in a ZEC are:

A. The use of biomass wood products, such as wood pellets or wood chips, in conjunction with a stove

B. The use of biomass products in a boiler operation, which then uses the heated water for thermal energy, or to drive a steam turbine

*I first became aware of biomass plants that would be applicable to ZECs during 2009 while I was working on the development team for a large sustainable residential and mixed-use community with the global development company Lend Lease Communities of Australia. They introduced KMW ENERGI, a privately owned company based in Norrtälje, Sweden. The company, which was founded in 1958, builds wood-fired boilers for wood processing, sawmill, pulp and paper industries. The technology is appropriate for hot water and steam for heating and other processes, power generation or combined heat and power, which produces electricity and thermal energy. (KMW ENERGI Inc. 2013)*

During 2010, I learned that Ameresco Inc., a leading independent provider of comprehensive energy efficiency and renewable energy solutions, built a biomass energy facility for the NREL campus in Golden, Colorado. The combined-heat and power (CHP) system is reported to work very efficiently. Robert Welch, Chief Technology Officer at TowardZero.org, explained his findings about the operation of the NREL biomass plant:

*The fuel supply for a biomass plant needs to be very consistent, and wood products must be free from dirt. Inefficiencies occurred whenever the fuel stocks were changed. Even though any of the biomass fuels could be effective, changing the type of fuel supply caused inefficiency in the biomass facility operation. (Welch 2012)*

Since that time, I have been involved with an initiative to clear wood killed by pine-beetle infestation in the Rocky Mountain forest region in order to reduce the risk of forest fires and to use that wood to provide biomass-heating solutions to ski resorts. In such instances, the accessibility of sites by trucks or railcars carrying wood chips to the boiler operation is paramount to the design. The configuration of roads, rail and railroad spurs is key to the feasibility of these types of ZEC projects, including those at ski resorts.

*Environmentalists worry about the upstream effects of biomass energy projects, citing forest management problems that may result. (Berwyn 2012)*

ZEC planners who will deploy biomass need to consider the source of their fuels, and the impacts to forests that may occur as a result. Most biofuel facilities do not incorporate electric generators, and therefore produce heat in a boiler operation. The use of that heat is an important consideration, as are fuel transportation methods and location of storage areas. Increasing use of biofuels will undoubtedly lead to a certification process for biomass fuel stocks in the future.

### Waste to Energy

Power generated from the incineration of waste is considered renewable energy in some states. Waste power plants may provide a renewable power source to a ZEC if located in the facility's proximity, if the power rate and cost align, and if the ZEC can find a means to legally transmit the power. Most waste-to-power plants are built when there is a power purchase agreement (PPA) between a waste company and local utility company, making the construction and operation of the waste to the electricity facility financially viable. The PPA will typically be for a term of twenty to thirty years, and when complete, the waste to power operator may sign another agreement to sell the power, in this case to a ZEC. A ZEC planning team should determine if there are any types of facilities like this in their local area, how the term of the PPAs aligns with the schedule for the ZEC development, and the generating capacity, reliability and cost of the power.

### The Future of Renewable Energy

Millions of grid-connected, off-grid homes and other buildings are being equipped with solar panels, wind generators, micro-hydro generators, geo-thermal ground loops and supplemental biomass heating systems to minimize their use of fossil fuels through increased use of renewable energy.

I liken the use of renewable energy to the adoption of the Internet. Initially, it was only the few who understood the promise of Internet technology. Then, everybody climbed on the bandwagon, and what was once rarified, became a global practice. Renewable energy has been through a similar first phase, demonstrating that it can work and that it can compete financially, and the trend has gathered a large base of early-adopters. It is becoming easier to understand which technologies work well in each sub-climate, where to buy them, who can fix them if they break down, and how to finance them to gain immediate savings and benefits. My sense is that mass-adoption is beginning to occur that will reduce prices of renewable energy equipment further, and soon everyone will have the equipment and be able to embrace renewable energy, much like they have the Internet.

The downsides of renewable energy are the up-front costs, the uncertainty of innovation, and the risk of obsolescence. It is also important to recognize the systemic challenge of balancing supply and demand.

> Although it is conceivable that behaviors, or what Lovins calls mindful use, or as the French say, sobriété, (Lovins, Chief Scientist and Founder, Rocky Mountain Institute 2013) could help match time-of-use patterns for energy demand with renewable supplies, this is not enough to level the supply/demand at every minute of the day. Renewable energy advocates promote a mix of energy types.
>
> In Hot, Crowded and Flat, Thomas Friedman paints a picture of how community members will use their smart phones to monitor and control their energy use, energy expenditures, EV storage and overall sustainability. (Freidman 2009)
>
> At scale, the collective Wisdom of Crowds (Suroweiki 2004) can play a material part in providing demand-side management (DSM) to balance the supply and utilization of energy.

Even after adding DSM and storage to a ZEC microgrid, there is still a need for energy to be dispatchable, providing the energy supply of last resort, to ensure the community energy supply is appropriately reliable. This is an area where the utility industry has a business opportunity to provide great value by providing "firming" energy to a ZEC's power supply.

Perhaps the biggest threat to renewable energy would be human apathy and resistance, these same behaviors that have caused many societies to crumble while progressive technology changes (like the Internet) have caused huge ground-shifts in world economies.

> *Renewable energy has even more potential to afford most people in this and future generations, a better life. According to the Rocky Mountain Institute, the five-billion dollars a day America will save by eliminating foreign oil imports will provide abundance in our national economy. (Lovins, Reinventing Fire: Bold Business Solutions for the New Energy Era 2011)*

> *How will a preponderance of renewable energy become available in America? Business is the only mechanism powerful enough to reverse environmental and social degradation. While it is true that individual activity is empowering …
> it cannot of itself change the nature of environmental and social degradation.
> (P. Hawkin 1993, 1999)*

### *Getting Renewable Energy to Work for Business*

I hope that the business of ZECs helps to drive economic and societal advancement for our entire civilization. The friction today is an obsolete power grid and the business structure of utility companies which are not incentivized to make the change to renewable energy and energy efficiency. This change is foundational, however, to widespread adoption of ZEC's.

Developing ZECs is a solid step forward, but there is also a need to bring the American utility companies along in the process, as Amory Lovins described:

> *Obviously if you have zero-net buildings, a volumetric tariff, cents per kilowatt-hour; it guarantees that the utility will receive zero net-revenue over the year. The RMI Electricity Lab (e-lab) has devised a new kind of tariff that tells the customers and utility how to compensate each other for the value exchanged. (Lovins, Chief Scientist and Founder, Rocky Mountain Institute 2013)*

# SECTION 10
# ELECTRIC VEHICLES

*According to the Rocky Mountain Institute, in the next two decades it will be increasingly more common for people in the US to travel to and from work in an electric vehicle (EV). This mode of transportation takes many forms: a bus, a van, or a car-share or car-pool vehicle that accelerates like a sports car, makes little engine noise, has no tailpipe emissions and is not reliant on fossil fuels for operation. The electricity used to power such EVs can come from fossil-free solar, wind, and other renewable sources. (Lovins, Reinventing Fire: Bold Business Solutions for the New Energy Era 2011)*

When I spoke with Lovins about EVs and ZECs, he pointed out synergies that go beyond the value of powering our cars with renewable energy, including the value for the Smart Grid:

*This is not only about getting off oil and using off-peak renewables better, but also better enabling the grid to integrate variable renewables. That turns out to be a very important benefit to the whole electricity system. So you really get all sectors renewably powered except industry... all buildings and cars. The all-electric vehicle fleet would represent an estimated fifteen percent (15%) of the community electricity system, and provide a controlled varying load with smart charging and discharging. (Lovins, Chief Scientist and Founder, Rocky Mountain Institute 2013)*

The first step toward such a future was the adoption of hybrid electric vehicles (HEV) that used conventional internal combustion engine (ICE) propulsion systems in combination with electric propulsion systems.

While initially perceived as unnecessary (when introduced near the beginning of this century) due to the low cost of gasoline, the last one hundred-plus years has proven that fossil fuels have a finite supply and carry tremendous environmental and health risks.

*Fast forward to the twenty-first century when worldwide increases in the price of petroleum caused many automakers to release hybrids in the late 2000s. HEVs are now perceived as a core segment of the automotive market. (Electric Drive Transportation Association 2013)*

More than 5.2 million HEVs were sold worldwide by 2012, led by Toyota Motor Company (TMC) and closely followed by Honda and Ford. This trend compelled automakers to introduce plug-in hybrid electric vehicles (PHEV) and electric vehicles (EV). General Motors launched the 2011 Chevrolet Volt series plug-in in December 2010... and Volkswagen and BMW will introduce EV passenger vehicles in 2013.

*The future holds much promise for EVs. In addition to battery-powered motors being commercialized today, early-stage vehicles that utilize onboard hydrogen fuel cells that generate electricity are being used experimentally (Whitcomb, Field Report: National Renewable Energy Center - Visitor Center 2012, Nov, 11). Lightweight HEV, PHEV and EV concepts were originally brought to the world's attention through the Rocky Mountain Institute Hypercar Center and are now*

> *poised to dominate vehicle manufacture before 2050. (P. L. Hawkin 1999) (Muller 2013) (Shepard 2013)*

Now, imagine that it is the end of the workday and you return to your own residence within a zero energy community, ready to plug-in your PHEV or EV for recharging. Like many of the residents in your ZEC, you will be doing this at approximately the same time each weekday. You and your neighbors will likely be grateful that your ZEC was planned to support the convenient plug-in vehicle (EV), complete with plug-in ports and/or inductive wireless charging stations. In addition to planning for the convenience of charging EVs, the Sustainable Homeowner's Association anticipated the electrical supply/demand at this hour of the day. This demand would have been figured into the originating goals of achieving a net balance of energy within the respective ZEC.

In a traditional neighborhood, there may be no provision for plugging in an EV, or the cycle of use that results in many drivers all plugging in at the same time presents stresses on the neighborhood's electrical grid that may cause a brownout or blackout.

I was skeptical about how a developer would react to the requirement for EV charging at a ZEC. Not seeing EVs in current practice, the developer would most likely question the projected need and legitimacy of this provision. Then, during 2012, Montgomery Force, Owner of Force Consulting and Executive Director at Lowry Redevelopment Authority said that planning and paying for EV power connections at their planned ZEC of 800 homes is not an issue. The Design Guidelines will require this on every home.

This scenario illustrates the advantage of bearing up-front costs of such future planning. These costs are ultimately marginal in the context of the community electrical system, compared to the cost and inconvenience of retrofitting a home or community that does not make the same provisions.

> *The effectiveness of EVs is not limited to issues of transportation only. Because EVs incorporate battery electric storage, an EV vehicle can power an entire household, according to Nissan advertisements for its LEAF model. (BBC 2012)*

This claim requires a more detailed disclosure, stating how big a home, how long will the power last, and then what to do when the car batteries are dead. It will be very interesting to see how quickly the acceptance of EVs will be popularized through advertising.

Using the storage capability of our electric vehicles to smooth the power supply of a building or community is an idea with merit. The integration of EVs in the electrical system could permit a better balancing of intermittent electricity supply from solar and wind generation compared with the actual minute-to-minute demand of the homes or other buildings in the ZEC.

> *Already the auto industry has introduced a range of EVs from the Nissan Leaf to the Tesla Model S, totaling six production models for sale in the US as of 2012. Dozens more are scheduled for introduction to the US market. (Lovins, Innovation in Efficient Vehicle Design and Energy Supply 2002)*

> *2013 will be a landmark year for electric vehicles—with BMW, among others, entering the EV market as the trailblazers of this market trend. Chevrolet, Ford, Nissan and Tesla will expand their offerings dramatically. By 2014, it is expected that almost every major worldwide car manufacturer will have significant plug-in electric cars in the market. (Muller 2013)*

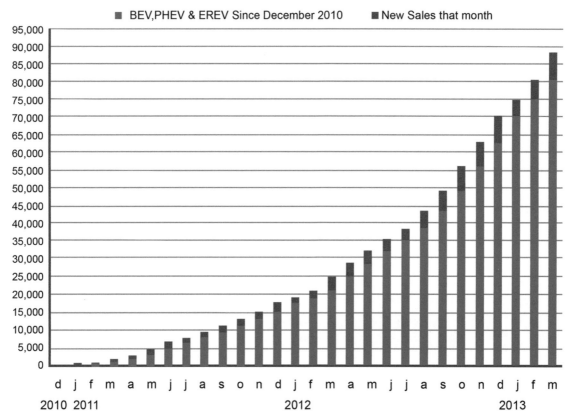

## CUMULATIVE U.S. PLUG-IN VEHICLE SALES

*Figure 18 - ELECTRIC DRIVE TRANSPORTATION ASSOCIATION (US MARKET) FOR ELECTRIC VEHICLE SALES. ALTHOUGH THE TOTAL NUMBER OF EVS SOLD IS VERY SMALL COMPARED TO SALES OF CONVENTIONALLY POWERED CARS, THE CUSTOMER UPTAKE RESEMBLES THE STEEP ADOPTION CURVE FOR INTERNET SERVICE FIRST SEEN IN THE EARLY 1990'S. (Used by permission of Electric Drive Transportation Association 2013)*

At a high level, the ZEC must consider its place in the larger environment and how access to, and through, the ZEC interacts with the ecological, urban and architectural design and planning of a large city or metropolitan area layout where the ZEC is contained. A ZEC can be planned in several ways that contribute to the conservation of transportation-related energy.

ZECs address two of the largest energy uses: transportation and buildings. By assisting the transition to electric vehicles, ZECs hold the potential to materially reduce the use of fossil fuels and provide huge economic, environmental and social benefits. Accommodating EVs with charging stations is an important virtue of ZECs.

> *Transitioning to EVs powered by renewable energy provides a direct path to de-fossilization of transportation energy, which is the largest domestic energy use. For all of these reasons, the National Renewable Energy Laboratory (NREL) includes the accommodation of EVs as a requirement in their Definition of a Zero Net Energy Community. (Carlisle, AIA, VanGeet and Pless 2009)*

The linkage between EVs and ZECs is one of the most important attributes of the ZEC. According to NREL, Japanese homebuilders are bundling EVs with their new homes. Imagine driving an electric car fueled by your ZEC's renewable energy.

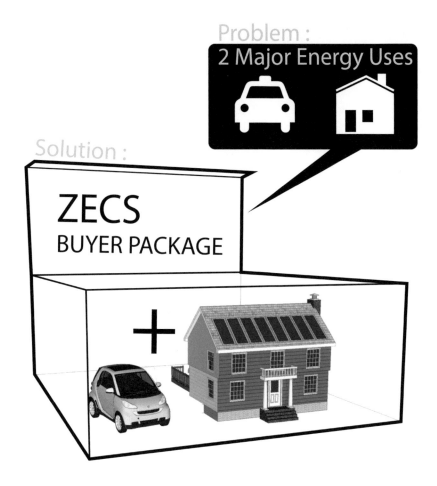

*Figure 19 - A ZEC SHOULD ACCOMMODATE THE CHARGING OF HIGH-EFFICIENCY ELECTRIC VEHICLES AS PART OF THE COMMUNITY AND EACH BUILDING'S ELECTRICAL SYSTEM DESIGN. IN ADDITION, A ZEC PLAN SHOULD ACCOMMODATE FOOT, BICYCLE AND MASS-TRANSIT TRANSPORTATION.*

# SECTION 11
# TRANSFER AND STORAGE OF ENERGY

Imagine a large factory that produces sixty new widgets every second and a fleet of trains and tractor-trailers lined up at the loading dock of the factory ready to carry those widgets to market. Predictably, there are perturbations in the rate of marketplace demand for those widgets, and often, many widgets must be stacked and stored in warehouses. In this case, storage acts as the bridge between supply and demand.

Now, imagine a large bakery that produces sixty new loaves of bread every second and a similar transportation system that carries those loaves to market. As with other goods, there are variances in the rate of demand for those baked goods in the marketplace, and often, many loaves are liquidated for pennies on the dollar before they become too stale to sell. In this case, storage cannot act as the bridge between supply and demand as in the previous example, and that is because the loaves will spoil much too quickly.

Imagine the factory that produces electricity instead of widgets or baked goods, and instead of conventional transport to deliver those goods to the marketplace, electrons are produced, and are delivered to buildings, plants and streetlights at a speed that is approximately one-half of the speed of light. The electricity, unlike the widgets and bread, is only suitable for immediate use.

The result of an oversupply or undersupply of electricity can cause a blackout or brownout. Oversupplies cause protective equipment to switch-off circuits, and brownouts can occur when there is undersupply.

Alternating current (AC) is a type of electrical power that is utilized by the power grids in the US, and the standards require that the electricity maintain precise voltage (V) and frequency (Hz) characteristics. Typically, 110, 120, 277 and 408 volt (AC) is used by retail customers of electricity. The voltage in electric power transmission lines used to deliver electricity from power plants can be several hundred times greater than consumer voltages, typically 110 to 1200 kV (AC). In all cases, sixty hertz (60 Hz) is the required frequency in the US.

> *When the supply and demand are not perfectly matched, the voltage, or frequency, may begin to drift outside of the allowable power quality tolerances. In the case of a traditional power plant, the ability to regulate the amount of power output is very slow reacting. Paul Thompson, a patent agent who specializes in energy transfer and storage, explained that adjusting the output level of large power plants requires as much as fifteen hours after the control systems call for that adjustment. (Thompson 2013)*

The power grid control centers manage the balance of supply-demand by forecasting resource demand and utilizing peaker plants that react more quickly than large base-load power plants to compensate and deploy various means of storing the electricity. There are predictable patterns that can be used to forecast the base power resource requirements according to time-of-day and weather conditions.

Additionally, the grid operators buy and sell power on a live spot market that allows the intake and offtake of power, as an additional means to optimize the balance of the voltage and frequency, and the economics of the utility.

The storage of electricity is helpful to prevent brownouts and blackouts, and if enough storage were available, the power plants would only need to provide the average amount of electricity required, instead of peak power. What are the options?

The term "storage" may be somewhat of a misnomer, because the electricity, to be stored, must be converted into something else. Batteries convert electricity into a chemical reaction. Pumped-water storage converts the electricity into a gravitational force. Ice-energy storage converts the electricity to frozen thermal mass. Compressed-air storage converts electricity to pressurized air.

Thompson points out that energy to be stored must survive several inefficient processes: In the case of a battery, alternating current (AC) power must first be converted to direct current (DC) by a convertor. Then the battery anode converts that electricity into a chemical reaction in the cells surrounded by electrolyte. Next, the battery cathode converts that energy back to electricity, which is then sent to an inverter, which converts the electricity back to alternating current. In each step in the process, the energy conversion is inefficient, and the overall inefficiency of the storage process is the cumulative result of those inefficiencies. Given these parasitic losses of efficiency, there is simply not a good way to store electricity.

> *Peter Kelly-Detwiler, a contributor to Forbes Magazine, interviewed Gary Wetzel, Director of Commercial and Industrial Business Development for S.C. Electric Company, about the prospects for storage technology evolving. Wetzel compared current-day storage technology with the solar power technology of twenty years ago when we saw solar deployed for off-grid applications, such as cabins, highway signs, offshore buoys, and calculators. Storage is being utilized [today] mostly in high-value, niche applications - but like many emerging technologies - looks to go mainstream in the long run. (Kelly-Detwiler 2013)*

Make no mistake, storage technology works today, and is deployed heavily for mission critical applications such as data storage, as well as being a part of the growing fleet of microgrid projects in the US. The critics of storage efficiency seem to have one foot in potential improvements, and the other foot in the current high cost of the technology. Would the efficiency be seen as a problem if the price were lower? I do not think so. I believe the dilemma with storage is less about technology and efficiency, and more about cost.

Industry interests in storage are driven by new uses of the technology. An increasing supply of intermittently operating solar and wind creates a growing problem with system imbalances throughout the nation's grids. Because solar voltaic panels produce direct current (DC), batteries are more complementary with solar because they are also DC. The electric vehicle industry is investing heavily in battery technologies for mobile applications, and that may produce solutions for stationary use, as well.

It is reasonable to assume that efficiencies will increase, size and weight will decrease, and cost will drop as the batteries of the future become market commodities. Today, entrepreneurs are experimenting with chemicals and business models.

Timothy Collins, Sr., CEO of KleenSpeed Technologies, a battery development company, envisions storage as a service. The chemicals do not matter, the production costs do not matter either the need is to be able to amortize the cost over the long-term, and sell storage as a service. This concept may yield many benefits.

Another way to look at electricity storage is through the eyes of economic opportunity. In 2001, I learned of an operation located in Ohio that purchased electricity from the grid. They purchased when rates were lowest (in the middle of the night). They then converted the electricity to compressed air stored in a limestone cavern—and then back to electricity during the middle of the day, when power costs were 400% higher than the purchase price. This sort of arbitrage is much more likely in the future.

The time-dependent value of energy is an important concept to embrace, because according to energy industry leaders, rates for electricity will have to vary at every moment of every day within the nineteen climate zones in the US. Today there are huge price variations between the middle of the night and the middle of the day. In some cases, electricity has a negative price at night when wind energy supplies are highest and demand is lowest.

The requirement for storage involves both the power level and the required storage time. Increase either and the cost goes up; decrease either and the cost goes down. Therefore, having enough storage capacity to stabilize an electrical system for a few minutes, allowing load demand to subside or additional generators enough time to start up, may be an inconsequential cost—while running an entire community for many hours would not be affordable to most.

### Common Types of Energy Storage

Batteries are direct current (DC) devices. Solar photovoltaic panels (PV) produce direct current, and this makes batteries compatible with photovoltaic panels. Batteries equipped with invertors and convertors are less efficient because parasitic losses of the system can reach seventy percent. Even though the efficiency is low, large battery stacks are used to stabilize micro grids and to provide continuity of operation in conjunction with emergency generators.

Pumped-water storage uses AC electrical power to operate an electric pump that forces water uphill to a reservoir and an electrical turbine generator that provides electricity when the water is released from the reservoir. The conversion is a very inefficient round-trip process; however, the quantity of energy that can be stored can be significant.

Salt-phase change storage is a method of thermal storage, which can indirectly benefit an electrical system because the thermal energy can displace the need to produce refrigeration at times of peak demand. This is because it is a product of phase change of a chemical process that is associated with matter changing states; e.g., liquid to gas or visa-versa.

Every ounce of matter cooled one degree provides one British Thermal Unit (BTU) of energy, and this is efficient. Ice-energy storage is based on this principle, and this technology is being deployed widely in California where electrical power is costly and environmental concerns are important. The utilities signal the ice-energy units to make ice when electrical supply is high compared to demand and to use that ice to cool buildings when demand is high compared to supply.

A common technology for short-term electrical storage uses a heavy flywheel connected to a motor and generator. When the electrical supply is interrupted, the flywheel continues to spin. Although this is mostly used in the Europe, I have seen flywheels in the US. When I visited the Skywalker Ranch production facility in 1995 (where George Lucas produces Star Wars movies), I saw a flywheel in use. The producers explained that they had power reliability problems caused by the fluctuation of power demand associated with the turning on and off of the milking machines at the neighboring dairy farms, and that they installed the flywheel, which then eliminated the problem.

Storage can improve electrical reliability. Storage can provide price arbitrage. Storage can improve frequency control in an electrical grid.

### *Consumers Who Adjust Their Use of Electricity Can Do All the Same Things*

Market mechanisms that constantly adjust the value of electricity, storage, and thermal energy conversions provide efficient means to balance electrical supply and demand. The yellow blinking light on your smartphone can signal the opportunity to save money by delaying your dishwasher's operation or the need to conserve ZEC power until 7:00 pm. By storing up individual needs for power, a consumer can be part of a storage solution.

Electric vehicle battery storage may be the best alternative for ZECs because the collective fleet would provide a significant storage utility, and the cost of other storage equipment purchases could be reduced or avoided completely.

> *I think that in the future, what we are going to see increasingly is a combination of solutions with Smart Grid self-healing capabilities. We'll see solar integrated with gas-turbines, fuel cells, wind, battery backup, and all of it linked with interconnected communications. The storage system is built to interact and optimize its behavior with the local power grid. A master controller utilizes rules-based algorithms to look at variable power rates and determine how much energy to store during the cheaper off-peak hours and how much to release into the market when prices are higher. (Kelly-Detwiler 2013)*

### *Conclusion*

Energy storage should concern a ZEC, especially if the ZEC is designed to operate independently of the regional electric grid (off-grid). Before implementing storage, which is almost always more costly than generation, measures should be taken to reduce peak load conditions in a ZEC, because that reduces the need for new energy generation assets and provides better utilization for all equipment, thus reducing cost.

Could people and technology synchronize their energy use patterns with the energy supply? Can they shave usage when demands are high? ZEC planners can develop innovative time-of-day pricing to regulate power through a market mechanism. To the extent they cannot, or will not, batteries and other technologies are needed to firm intermittent renewable energy sources and smooth generation requirements regarding peak loads.

Harnessing the storage capacity of plug-in electric vehicles in the ZEC may be the most effective as asset utilization goes up, and community members are encouraged to use EVs because they do not require fossil fuels or generate significant pollution. EVs parked throughout our communities, theoretically, can operate as energy storage devices when they are parked. ZEC planners can develop innovative means to compensate EV owners for the use of their storage.

# SECTION 12

## HYBRIDIZATION

*Hybridization of energy refers to combining energy generation sources, which can blend renewable and a single non-renewable energy, or two or more sources of renewable energy. Hybridization is the technique required to realize the aforementioned "all-of-the-above" strategy referenced by President Barack Obama in his 2012 State of the Union address. This all-of-the-above energy strategy underlies America's ability to transition to utilizing a blend of renewable energy as our primary energy source. (Lovins, Reinventing Fire: Bold Business Solutions for the New Energy Era 2011)*

The hybrid energy model is important to zero energy communities for two reasons:

(1.) The cumulative power from multiple renewable energy supplies may be required to meet the zero-net balance of energy sources and energy uses.

(2.) A ZEC that operates off-grid may need to hybridize renewable energy with dispatchable non-renewable energy sources, such as a generators, in order to make its energy supply reliable.

The Montreux Clean Energy Roundtable was a conference of approximately fifty energy experts held in Aspen, Colorado, during June of 2012. Representatives of government, institutions, and executives from many large companies (including major oil and gas companies) discussed hybridization of energy for two days.

*Amory Lovins was the keynote speaker. In his book Reinventing Fire (2011), he described the use of many energy sources, dramatic improvements possible in the efficiency of energy uses, and how his proposal for a hybrid approach would replace America's hydrocarbon energy with renewable energy by 2050. (Montreux Energy 2012) (Lovins, Chief Scientist and Founder, Rocky Mountain Institute 2013)*

Presentations during the conference demonstrated a plan for America to utilize increasing amounts of natural gas and to bridge gaps in the effectiveness of renewable energy—those that result in intermittent renewable energy from solar and wind and tidal power. Hybridization provides a definite pathway to de-fossilize energy in America, and zero energy communities can serve as a platform to drive this trend toward utilizing renewable energy for transportation, buildings and industry. Combining multiple energy sources and alternative fuels includes many possibilities, as does the exploration of the behavior of microgrids and how people increasingly interact with their energy systems.

*Brandon Owen, the electricity market strategist at the General Electric Company presented GE products that are designed with in-built hybrid capabilities. These agile generators synchronize with the electricity grid to fill gaps in demand caused by gaps in the supply of intermittent renewables from wind, solar, or tidal energy. Batteries are used to bridge the short generator ramp-up time. (Owen 2012)*

*A Honeywell International (HON) representative presented their microgrid intelligence, controls, and smart building capabilities. I was already familiar with some of HON's material because I had led the architecture and engineering team for India's smartest building during 2009-12. (Whitcomb, Bharti-Airtel Network Experience Center 2012)*

Honeywell had provided the intelligent building management system (iBMS) for that building. At the conference, I learned even more about HON's developing capability to manage large grids.

At an off-grid ZEC, where there is no steady supply of power available, ZEC planners will likely need to deploy generator(s) that can firm up the supply of energy when solar and wind are not available because of settling wind or clouds obscuring solar energy. Even though generators start up and ramp to full power much more quickly, there will be a need for short-term storage to allow the generators enough time to ramp up to full power.

If a ZEC had storage that would last for a month without wind or sun, it could eliminate the need for engines like the ones GE manufactures. Unfortunately, battery storage is expensive, and when you increase the timeline for the storage supply, the cost rises incrementally. A long-term storage for a ZEC would have huge costs, only affordable by multi-millionaires. However, a few minutes of energy supply through batteries cost much less and can supply the community's power while generators start up.

*Many of these generators accept multiple fuels, including biofuels and natural gas. The supply of renewable biomass or biofuels to operate a generator would be ideal. If, for any reason, renewable fuels are not available at reasonable cost, natural gas can assure continued energy convenience and safety to the community. In addition to being reasonably assured in supply, and with predictable cost, natural gas is the least polluting fossil fuel available today: (NaturalGas.org 1998)*

## FOSSIL FUEL EMISSION LEVELS
POUNDS PER BILLION BTU OF ENERGY INPUT

| POLLUTANT | NATURAL GAS | OIL | COAL |
|---|---|---|---|
| CARBON DIOXIDE | 117,000 | 164,000 | 208,000 |
| CARBON MONOXIDE | 40 | 33 | 208 |
| NITROGEN OXIDES | 92 | 448 | 457 |
| SULFUR DIOXIDE | 1 | 1,122 | 2,591 |
| PARTICULATES | 7 | 84 | 2,744 |
| MERCURY | 0.000 | 0.007 | 0.016 |

*Figure 20 - INCREASED ESTIMATES REGARDING THE DOMESTIC RESERVES OF NATURAL GAS, COMBINED WITH SIGNIFICANTLY LOWER EMISSIONS, MAKE IT A CLEAR CHOICE OF FUEL FOR NON-RENEWABLE ENERGY. (EIA - Natural Gas Issues and Trends 1998)*

A typical hybrid energy configuration may include micro-turbine or aero-derivative gas turbines to firm solar and wind, bridged with battery storage, and controlled by micro grid software to produce reliable electricity. Another configuration is to deploy a steam turbine and boiler that use a combination of gas, liquid biofuel and/or biomass for fuel to create steam.

There are, of course, time-of-use considerations with all these energy technologies, though that needn't prove a liability. Solar energy is a very useful source of energy to power air conditioning, while wind that blows at night and coincides with the charging of electric vehicles at nighttime is another example of efficient time-of-use. A symbiotic relationship between intermittent sources of wind and sun can invite creativity into how hybridization is achieved. When the emerging storage services and electric vehicle technologies are more of a mainstay, a ZEC could share battery services with electric vehicles parked throughout the ZEC. A large number of parked vehicles in the area of a ZEC microgrid would improve the reliability and power quality and possibly reduce the ZEC's renewable energy requirements.

This microgrid market is experiencing considerable innovation and growth, and this suggests that the technology is reliable and cost-effective today. The industry is gaining experience with the growing number of microgrids, and these are all hybrid energy solutions.

Navigant Consulting has tracked and reported the trend toward hybridization in their coverage of microgrids and reported dynamic market growth that is good evidence that ZEC micro-grids are growing. As of their 2Q 2013 report, Navigant Consulting has identified:

> *A total of 3,793 MW of total microgrid capacity throughout the world is up from 3,179 MW in the previous tracker update in 4Q 2012. North America is still the world's leading market for microgrids, with a planned, proposed, and deployed capacity of 2,505 MW, representing 66% of the global capacity. This represents an additional 55 projects since the 4Q 2012 update, and an additional 417 MW. Of the total North American microgrid capacity, 1,459 MW is currently online and over 1,122 MW is in the planned/under development or proposed phase. (Navigant Research Consulting - Asmus 2013)*

> *ZEC planners should contemplate how their decisions may impact environmental and health issues associated with natural gas, increasingly extracted through hydraulic fracture drilling, or "fracking." The gas industry requires modern regulations requiring greater environmental protection by all gas operators. (Montreux Energy 2012)*

Hybridization provides the ability to transition to renewables gradually, with the help of non-renewable and renewable fuels. Hybrid blending means that new renewable projects can more easily come alongside existing legacy energy solutions to supplement and support a transition. Increasing the use of renewables can be an incremental process, and the value and benefit of hybridization deserves ample attention by ZEC planners.

The ramifications of hybridization are tremendous. Hybridization techniques challenge the idea that there would be a switch from non-renewable energy to renewable energy, and instead suggest that there is a blending of the two. Blending renewable and non-renewable initially challenged my own all-green ideology, causing me to think of moving *toward* green instead of *to* green. For this reason, I chose to name my non-profit foundation TowardZero.org (TZO), and to encourage business and government to move large projects in the right direction toward net zero, versus pursuing ideological perfection. This type of transition appears easier and may pay more green dividends ultimately when America approaches the point of a fifty-fifty percentage for renewable and non-renewable energy. TZO believes that such an approach applied on a wide scale trumps a handful of projects that succeed in implementing 100% renewable energy.

Just as alloys perfect the attributes of several metals, hybrid energy makes the best use of many energy components. ZEC developers should consider hybridization when a ZEC is being planned. The following are recommendations:

ZEC planners should investigate all energy sources available, and consider using multiple sources. When evaluating intermittent energy sources, consider the time of day the energy is produced, and any seasonality associated with those energy sources. This is called a "site screening" and can be performed utilizing NREL's regional data about renewable energy.

ZEC planners should also consider the energy uses. How much electricity is needed to power homes, buildings, and exterior lighting? These energy uses must be evaluated for each time of the day, and for seasonal variations. In addition, what is the requirement for emergency power to support mission-critical telecommunications, data centers, first responder and medical facilities located in the ZEC?

After considering the sources and uses of energy at each time of day and season, the ZEC planners can assess effectively the hybridization of multiple sources of energy to match the availability of energy for their given uses. This exercise can indicate what dispatchable energy generation (e.g., from an emergency generator) may be required to firm the supply. This exercise helps to make renewable energy practical. Hybridization is particularly important to an off-grid ZEC that must serve the load of the community reliably. The control software used within the ZEC may favor renewable energy and utilize storage to leverage renewable power even further.

Local communities are examining their energy plans in light of climate-change concerns. David Driskell, Executive Director, Community Planning and Sustainability for the City of Boulder, Colorado oversees many sustainability activities, and his city is considering new business models for their electrical utility in order to meet their climate-initiative. He commented about planning ZECs:

> The ZEC Guide puts the pieces together in a more comprehensive way than we had. Boulder is exploring the operational and financial tools you would have available to you as a retail utility, versus trying to engage with a private utility that is not that interested. We could be at 50% greener than we are today with [grid] price parity. (Driskell 2013)

# SECTION 13
# CONTROL SYSTEMS

Control systems are used to communicate with, and remotely control, the ZEC electrical and energy-related equipment systems. The system operator uses buttons, interactive voice response, touch, gesture, biometric, website, Smartphone and social media applications to enhance its ability to control the system and all energy sources and uses while monitoring the detailed attributes regarding energy utilization. At the same time, the system continuously records data about system events.

A ZEC control system monitors, controls and provides visualization and automation of the ZEC electrical system health, grid inter-operation, maintenance activity, safety, electricity rate, time-of-day, and reliability, including many types of data and reports. Control systems signal equipment to tell it to turn on or off, send a report, start or stop, speed or slow, switch channels, route, control volume, protect equipment, reset breakers and many other operations. The two-way communications allow control systems to collect data from connected equipment. Control systems can typically aggregate a number of data points and provide reports about the health of a ZEC system about equipment performance, historical data, and the need for maintenance.

Information generated may include technical characteristics, with data about the ZEC's voltage, current, frequency, temperature, harmonics, noise level, historical usage and billing data. Similarly, the data could include information regarding site weather conditions, solar generation, wind behavior, district heating system temperature, or any abstraction of that data. That information, in turn, can automatically trigger other commands by the control system or cause signal alarms.

Control systems cover a broad expanse of equipment communication and controller devices that have a wide range of sizes and complexity ranging from a small hand-held garage door operator to a large system control and data acquisition system (SCADA) for an entire metropolitan area. These control and acquisition systems can all be interfaced with each other to provide new system applications.

A ZEC homeowner's control system would report the household electricity use and costs and would facilitate the household in programming the smart house's automated operation of primary climate control equipment, lighting levels, car charging, smart appliances and home theater controls. This would create a script for the systems to behave according to the system user's preferences and provide smart house operation of the following:

- Primary equipment for controlling temperature, humidity, ventilation, pumping and sprinkler irrigation
- Lighting controls, including zones, dim-level, colors, daylight control, energy conservation
- Appliances, including oven, dishwasher, refrigerator, washer/dryer , computer and home theater equipment (channel, volume, record)

The average American interacts with hundreds of control systems each day. We are well aware of the controls we use to program our alarm clock, turn on lights, heat the house, automate the coffee pot, open a garage door, operate sprinklers in our yard, or record an episode of a favorite television program. At the grocery store, the doors automatically open for us and the cash register provides us with discounts for entering our "favored shopper" identification number scanned at the payment terminal. We may be less aware of the control systems that operate streetlights, traffic signals and building energy systems.

### *Benefits of Control Systems*

(1.)  Equipment can be remotely operated and monitored.

(2.)  Equipment operation may be automated.

(3.)  System reliability may be improved by bypassing failed equipment and expediting repairs.

Electrical management systems (EMS) manage grid functions, assuring proper voltage and frequency—including power plant dispatching, transmission control and electrical distribution system management. The system may include Smart Grid automation. This type of system is operated by most utilities in their control rooms.

Smart Grids overlap all control systems, forming a central database that publishes information to other systems and system users. The systems utilize two-way communications and can gather information that the control system utilizes to more intelligently execute programmed operation through software that is required to monitor and control the supply and demand of energy for all users—plus, assure that the output voltage throughout the connected grid is within tolerance.

Smart Grids will even monitor near-term and long-term climatic conditions so that the system is prepared for changes in weather and the smoothing of energy supply and use, accordingly. The incorporation of smart appliances into the Smart Grid provides a mechanism for peak-shaving.

The Smart Grid concept provides consumers with information and options to improve energy efficiency, utilize renewable energy sources and support social needs of the community. The Smart Grid is also the platform for defining how we will use and pay for electricity. National Grid, which operates in the US and UK explains:

> *A Smart Grid is an "intelligent" electricity distribution network that uses two-way communications, advanced sensors and controls, advanced meters, and computers that can help reduce customers' energy use, improve the efficiency and reliability of the electricity grid, facilitate the connection of distributed generation facilities to the system, and optimize the integration of renewable energy systems. It also includes in-home energy management systems and intelligent controls in appliances, giving consumers more choice and control over how and when electricity is used, which can save money and help National Grid operate its electricity network more efficiently and reliably for the benefit of all its customers. We expect the Smart Grid to play an important role in reducing greenhouse gas emissions and other pollutants, especially in how it can facilitate the connection of large amounts of renewable energy—more than is possible with the current electricity distribution system. It is also expected to be an important enabler for electric vehicles and plug-in hybrid vehicles, two promising technologies that can help dramatically reduce oil consumption. (National Grid 2013)*
>
> *Sunil Cherian, CEO of Spirea, a leading maker of Smart Grid software,*

*describes it as a Smart Grid that is a marketplace empowered by technology. He*
*stated that consumers would interact with the Smart Grid when they achieve*
*some payback relative to their energy cost. (Cherian 2013)*

Microgrids (a ZEC is essentially a microgrid) are a component system within the larger Smart Grid, with smart metering and features that allow users to input their preferences regarding their use of energy and their preferred notification and monitoring scheme. These systems provide electrical distribution system management, including regulating system voltage and frequency. These same systems may also dispatch back-up generators or control storage devices. This type of system may be used by campuses or regional systems that are microgrids, but not net zero-energy oriented. This software is often used by operators who work in control rooms.

Building Management Systems (BMS) are used to integrate management of lighting, energy and security within large buildings or campuses. These systems provide their operators with the capability to drill down into the details of electrical, HVAC, and video surveillance and security access controls, and they provide manual and automated operation.

Home Energy Control and Automation are sophisticated residential control systems that can help us to record our energy preferences and report ZEC energy trends and status. Control systems can optimize conditions among many equipment systems simultaneously. These systems can interface with one's television, smartphone and electronic signage installed in community spaces. These systems may interface with billing data for electricity, thermal, water and other community services—and even call for maintenance service.

There are countless examples demonstrating how the Smart Grid could turn off refrigerators for a few minutes each hour during a peak load—or warm or cool a home or office in advance of a pending outdoor temperature change. For example, a homeowner may use solar to charge an electric car prior to the arrival of forecasted clouds that will interfere with electrical generation by their house's solar panels. Users will access the Smart Grid-connected appliances and homes through TVs, computers and mobile devices where they can monitor their usage and establish their automation preferences.

Security plans should include securely managing control systems and information. This includes both physical security and cyber security. A combination of security and privacy considerations goes into determining who is authorized to control various functions of equipment, access data, publish information or provide notifications.

Data privacy issues come into play when one considers managing the privacy of individuals within the Smart Grid. This a key concern of consumers, many of whom consider data about their use of energy and related financial matters to be private. ZEC system users may prefer that some data be made public and other data remain anonymous. Smart homes are those that have the capability to integrate control of lighting, temperature, media equipment and appliance control intelligently according to time of day, occupancy status, exterior climate conditions, and user preferences regarding the use of energy and the price and the carbon content of the electricity. Many Smart Homes operating within the Smart Grid work together to balance costs, environmental consequences, and power reliability for a community. The following chart presents an example of how a home control system might manage preferences for the temperature and energy efficiency:

| CONTROL FEATURE | CRITERIA |
|---|---|
| Target temperature | 72 degrees Fahrenheit |
| Setback temperature | 67 degrees winter<br>77 degrees summer |
| Day of week /<br>Time of day | Weekdays between 6:00 am and 9:00 am<br>Weekdays between 4:00 pm and 10:30 pm<br>Weekend days between 6:00 am and midnight |
| Power condition | Local renewable energy is available |
| Power price | Less than ten cents per kilowatt-hour ($0.10 per kWh), |
| Exceptions | Window or outside door is open<br>ZEC is signaling peak load risk condition<br>Room unoccupied for fifteen minutes (15-min.) - except when ESPN network is showing |
| Linked commands | Turn-off the media system when unoccupied<br>Close curtains when exterior temperature exceeds interior temperature |

*Figure 21 - ZEC RESIDENTS HAVE THE OPPORTUNITY TO PROGRAM HOME ENERGY USE ACCORDING TO TIME-OF-DAY, PRICE, OR BASED ON THE CHARGE STATE OF THEIR EV—AND EVEN THE WEATHER FORECAST.*

# OTHER PROGRAMS AND STANDARDS

The program for zero energy communities (ZECs) may result from a group's desire to achieve broad sustainability objectives, where net-zero energy is not the single goal. By utilizing a combination of complementary programs or standards, and running them in parallel to the ZEC program, ZEC organizers may seek to accomplish additional standards for sustainability, city design, and/or energy and environmental goals, including green building.

Cooperation among ZECs can extend across the country, and other programs and standards can be shaped to work when those programs are blended together. Many green building and infrastructure standards are compatible with ZEC planning and development. ZEC planners may choose to specify established green building and standards as part of their conservation goals for the ZEC.

A broad vision and the goal to transform the shape of a human settlement through eco-city development may also incorporate the requirement for a ZEC. Such a development can incorporate many other requirements in addition to the goal of a zero-energy community solution. Author Richard Register speaks about "eco-cities" (an eco-city is a city built off the principles of living within the means of the environment):

*There may be opportunities for wonderful new architecture, transit solutions, and new approaches to public and natural open spaces. Therefore, the layout of a neighborhood, village or city incorporates consideration of environment, economics, community and art, as well as energy, and these all may overlap and influence the energy solution and the living experience within that community. The ideal "eco-city" has been described as a city that fulfills the following requirements: (Register 2012)*

(1.) Operates on a self-contained economy; resources needed are found locally

(2.) Has completely carbon-neutral and renewable energy production

(3.) Has a well-planned city layout and public transportation system that prioritizes transportation methods as follows: walking first, then cycling, and then public transportation

(4.) Maximizes resource conservation, including the efficiency of water and energy resources, waste management systems and recycling

(5.) Restores environmentally damaged urban areas

(6.) Ensures decent and affordable housing for all socio-economic and ethnic groups and improved job opportunities for disadvantaged groups, such as women, minorities, and the disabled

(7.) Supports local agriculture and produce

(8.) Promotes voluntary simplicity in lifestyle choices, decreasing material consumption, and increasing awareness of environmental and sustainability issues

*In addition to these initial requirements, the city design must be able to grow and evolve as the population grows and the needs of the population change. (Graedel 2011)*

*The vision for eco-cities is growing today as evidenced by increasing literature on the subject (Fiona 2010) and even prototype developments like the Tianjin Eco City, which is a Futuristic Green Landscape for 350,000 residents. The development is located in China and is said to include Sustainable Design Innovation, Eco Architecture, and Green Building by Inhabitat, which also says that it is scheduled to be completed in 2020. (Yoneda 2011)*

Eco-city goals are especially important when taking into consideration infrastructure designs, such as for water systems, power lines, etc. These must be built in such a way that they are easy to modernize (as opposed to the dominant current strategy of placing them underground, and therefore making them highly inaccessible.)

Also important to the the development of all human settlements are the social impacts of the projects and the socio-economic impacts. Who is to be served by the project? How can that program be made to be just and equitable in the friction context?

Where does the ZEC fit among other programs? The hierarchy of sustainability programs fits into these basic groups:

(1.) Global sustainability doctrines, charters and protocols that address global civilization, human settlement, natural resources, and climate change. Examples:

A. *The Brundtland Report, also known as Our Common Future, from the United Nations World Commission on Environment and Development (WCED) was published in 1987 in response to the United Nation's General Assembly's realization that deterioration of the human environment and natural resources required nations to rally together in pursuit of sustainable development. (World Commission on Environment and Development 1987)*

B. *The Kyoto Protocol is an international treaty that sets binding obligations on industrialized countries to reduce emissions of greenhouse gases. (United Nations Framework Convention on Climate Change 1998)*

C. *The Earth Charter is a declaration of fundamental ethical principles for building a jut, sustainable and peaceful global society in the 21st century. (The Earth Charter Initiative 1997)*

D. *The United Nations UN-Habitat deploys the goals of the Human Settlements Programme and the Habitat Agenda, which define the needs and plans for sustainable global development. (UN-Habitat 1997)*

(2.) Sustainable urban planning, and eco-city and neighborhood design programs and standards are aimed at initiatives responding to climate change, carbon reduction, water conservation and natural restoration. Examples:

A. Eco-city design principles

B. Envision infrastructure sustainability rating system: Institute of Sustainable Infrastructue (ISI)

C. LEED Neighborhood Design, a program of the United States Green Building Council (LEED denotes Leadership in Environmental and Energy Design)

D. *Living Building Challenge, the International Living Future Institute (ILFI) philosophy and advocacy platform, and certification tool. The institute offers green building and infrastructure solutions that move across scales (from single-*

*room renovations to neighborhoods or whole cities). They are the parent organization for Cascadia Green Building Council that serves Alaska, British Columbia, Washington and Oregon, and are affiliated with the United States' and Canada's Green Building Councils. (International Living Future Institute 2010)*

E. The ZEC that addresses community energy and environment, including electric vehicles, green buildings, renewable energy and behaviors

(3.) Green building and energy and environmental programs that address sustainable infrastructure practices and provide conservation of natural resources, including:

A. LEED Rating system by USGBC

B. Passive house design (PHIUS) by the Passive House Institute of the US

C. Household Energy Rating System (HERS) - State of California

D. Water: USGBC-LEED ND, ILBI and Envision

E. Waste: USGBC, ILBI and Envision

(4.) *Local organizations: It is recommended that ZEC planners reach out and discuss common purposes and resource sharing with organizations having jurisdiction or interests in common with the ZEC: (Bobrow-Williams 2013)*

A. State, county, city government officials

B. State energy offices

C. State Public Utility Commissions (PUC)

D. Solar, wind, and geothermal energy organizations - national and state

E. Local utility companies

F. Local water departments and water boards

G. Sustainability groups

H. Environmental groups (e.g., Sierra Club)

I. Parks and recreation authorities - local and state

J. Planning and development agencies

K. Local building departments

L. Fire inspector and permitting authorities

M. Economic development agencies - state, regional and local

N. Local Chambers of Commerce

O. Health agencies

P. Labor, employment and job training

Q. Low income housing and accommodation

R. State Departments of Transportation (DOT)

S. Mass Transit Authority - local

T. Rideshare and car share programs - local

U. Local community garden program(s)

V. Trade development organizations (e.g., National Council of Non-profits)

Planners should leverage programs that are already in place inside and outside of their community. Checking with the above-listed groups may help to indicate other synergistic programs and people. (See "Resources" at the end of Sections 20, 21, 24 and List of Community Energy Projects in the Appendix for addresses for many websites)

### *Diagram of ZEC Development Hierarchy*

A ZEC is compatible with a host of other place-making programs. From the top down: (1) global programs for sustainability join with (2) ZEC initiatives, (3) energy, environment and green building programs, and (4) economic development and community development programs:

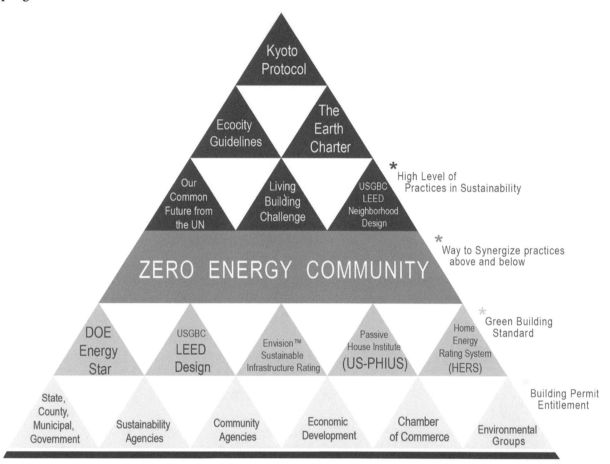

*Figure 22 - ZECS ARE SYNERGISTIC WITH A HOST OF OTHER DEVELOPMENT GUIDELINES AND PROGRAMS.*

**Real Estate Industry**

Because ZECs involve real estate, a ZEC strategy must also fit into the framework of the real estate development business and professional community. There are significant business and regulatory structures in place regarding development, construction, demolition, renovation, and related regulations, including entitlements and permits.

Real estate transactions involve specific procedures, contractual methods, and financial and insurance considerations. The industry utilizes very specific language that includes many definitions, terms-of-art, and jargon used only in the real estate industry.

The body of knowledge held by the local developers, real estate investors, bankers, insurers, title companies, brokers, appraisers, and regulators would be very difficult for a ZEC planning team to learn, and ZEC planners could realize a great benefit if they included real estate industry experts on their planning team. In addition, because states regulate this industry through regulatory authorities that enforce government and consumer interests in property transactions, financing, and insurance, a real estate attorney may provide considerable benefit to the ZEC team.

Opportunity: synergy to create positive, mutually beneficial results. Where do the organizations align in their objectives and missions? What would be the bilateral impacts to both organizations?

### Introduction to Several Green Building and Sustainability Programs:
### Envision Rating System (ISI)

The Institute for Sustainable infrastructure (ISI) - Envision Rating System was introduced in 2011. I spoke about it with Arthur Hirsch, Owner at Terra Logic, a sustainability-engineering consultant who was involved in the early application of the Envision for a wide range of projects in Colorado. His work applied the system to new projects involving storm water drainage, a tunnel restoration, a transportation project, and a recycling facility. He explained the background of the system:

> The system addresses all types of infrastructure projects other than buildings, references the Brundtland Commission's (Wikipedia 1983) definition of sustainability, and goes beyond that definition of sustainability to incorporate consideration for the people of the future. The Envision approach to infrastructure projects has taken on new considerations of long life, global warming, resilience to climate change impacts, embedded energy of materials, green procurement certifications, and removal and replacement. The involvement and collaboration with the community focuses on quality of life issues like water, sanitary services, noise and transportation. Envision does a good job of rewarding avoidance, but it goes beyond and rewards everything you do to enhance or restore the existing. The levels of achievement include improved, enhanced, conserving, and restorative, and the scoring echelons are similar to LEED Bronze, Silver, Gold and Platinum. (Hirsch 2012)

> The Envision system for rating infrastructure development (roads, bridges, water/sewer, campus physical plant, etc.) is the product of a joint collaboration between the Zofnass Program for Sustainable Infrastructure at the Harvard University Graduate School of Design and the Institute for Sustainable Infrastructure. Envision provides a holistic framework for evaluating and rating the community, environmental and economic benefits of all types and sizes of infrastructure projects. The Envision Rating System evaluates, grades, and gives recognition to infrastructure projects that use transformational, collaborative approaches to assess the sustainability indicators over the course of the project's life-cycle. (Institute for Sustainable Infrastructure 2012)

### USGBC LEED Neighborhood Design (LEED ND)

During 2009, the USGBC launched the Neighborhood Design (ND) program and today, neighborhoods, military bases, and other campuses are being certified under ND. ZEC planners will discover many benefits of LEED ND. (United States Green Building Council 2012)

Unlike other LEED rating systems, which focus primarily on green building practices and offer only a few credits for site selection and design, LEED for Neighborhood Development places emphasis on the site selection, design, and construction elements that bring buildings and infrastructure together into a neighborhood and relate the neighborhood to its landscape, as well as its local and regional context. The work of the LEED-ND core committee, made up of representatives from all three partner organizations, has been guided by sources such as the Smart Growth Network's ten principles of smart growth, the charter of the Congress for the New Urbanism, and other LEED rating systems. LEED for Neighborhood Development creates a label, as well as guidelines for both decision-making and development, to provide an incentive for better location, design, and construction of new residential, commercial, and mixed-use developments.

### USGBC LEED Green Building Design (LEED)

Zero energy communities closely share many features with the United States Green Building Council's (USGBC's) LEED, or Leadership in Energy and Environmental Design program, which provides building owners and operators a framework for identifying and implementing practical and measurable green building design, neighborhood and campus design, construction, operations, and maintenance solutions.

> *LEED has transformed—and is continuing to transform—the way built environments are designed, constructed, and operated. Nearly nine-billion square feet of building space has been LEED-registered and nearly two-billion square feet of building space has been LEED-certified since it was introduced in 2000. The LEED portfolio includes LEED Bronze, Silver, Gold and Platinum levels, each being more progressively sustainable. (LEED-certified Building Stock Swells to Two-Billion Square Feet Worldwide 2012)*

> *Now, LEED certification to a Gold level is obtainable at near the same cost as a non-LEED rated building, according to John M. Prosser, Professor Emeritus (ret.), School of Architecture / Urban Design, University of Colorado. Prosser attributes this to (1) the construction industry's familiarity with LEED processes gained over more than a decade of experience with the standards, (2) the availability of off-the-shelf green building components, which have come down in costs as they become more commonplace, and (3) the efficiency of the systemic construction process embodied in LEED. (Prosser, 2012)*

> *The United States Center for Disease Control (CDC) also provides a health-focused examination and acknowledgement of LEED ND's attributes. (Centers for Disease Control and Prevention 2010)*

### US Department of Energy - Energy Star Program

The United States Department of Energy also provides a bevy of sustainable resources and certifications, including Energy Star, that address buildings, heating and cooling, appliances and electronics at many levels affecting design and operational characteristics. In addition to rating the efficiency of appliances and electronics, Energy Star programs rate building energy efficiency and award certification as an Energy Star III building. This certification is applicable to commercial or institutional buildings and homes.

### California - Household Energy Rating System (HERS)

*The California Energy Commission established a home energy-rating program for residential dwellings as of 1999. This program established the requirements for field verification and diagnostic testing services used to verify compliance with California's Title 24, Building Energy Efficiency Standards. The 2009 update to the HERS regulations included the California whole-house home energy ratings requirements. These ratings apply to existing and newly constructed residential buildings, including single-family homes and multi-family buildings of three stories or less. The HERS system, which applies to homes, but not to commercial or institutional buildings, is also being used by many developers and builders outside of California today. (California Energy Commission 1999)*

### National Association of Home Builders (NAHB) ICC-700

National Green Building Standard, an American National Standards Institute (ANSI) approved standard is a nationally recognizable standard for Green Building.

### Passive House Institute (PHI)

*During 2012, the Department of Energy announced a partnership with Passive House Institute US (PHIUS) to certify Zero Energy Homes (ZEH) under the Builder's Challenge Program. Since 2008, the Builders Challenge Program has recognized hundreds of leading builders for achievements in energy efficiency— resulting in over 13,000 energy-efficient homes and millions of dollars in energy savings. Among these certified homes, more than 1,350 are considered net-zero energy-ready homes based on HERS scores of 55 or lower. The Energy Department's Challenge Home Program certifies homes that are 40% to 50% more energy-efficient than typical homes, while also helping to minimize the risk of indoor air quality problems and ensuring compatibility with renewable energy systems. (Energy Department's Office of Energy Efficiency and Renewable Energy 2012)*

### Local Building Codes

Land development and the construction and renovation of buildings is tightly controlled by a series of building codes that are determined by the local authorities having jurisdiction over the locality where development activities will occur. National building codes, electrical codes and fire codes are generally augmented in each jurisdiction.

Building codes may incorporate energy efficiency requirements, such as the California Title 24 code that influences the design of commercial and residential buildings and also addresses exterior lighting and other equipment. Like the California Household Rating System (HERS), these standards may be adopted by authorities and commercial users for projects built or operated in other states. Consult local architects and code consultants concerning the exact requirements for the ZEC being planned.

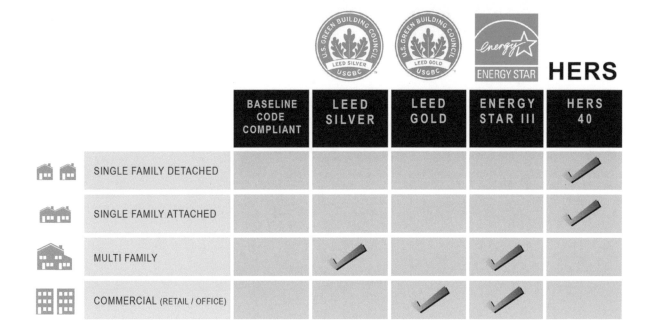

*Figure 23 - ZEC PLANNERS UTILIZE ENERGY EFFICIENCY AND SUSTAINABILITY PROGRAMS SPECIFIC TO BUILDING TYPES. (Courtesy of Design Workshop and Lowry Redevelopment Authority)*

The applicability of standards varies based upon whether the target project is a city, neighborhood, campus or commercial or residential building. The charts above and below helped the planning group at Lowry's ZEC to better comprehend the applicability of codes to building types being planned in that location.

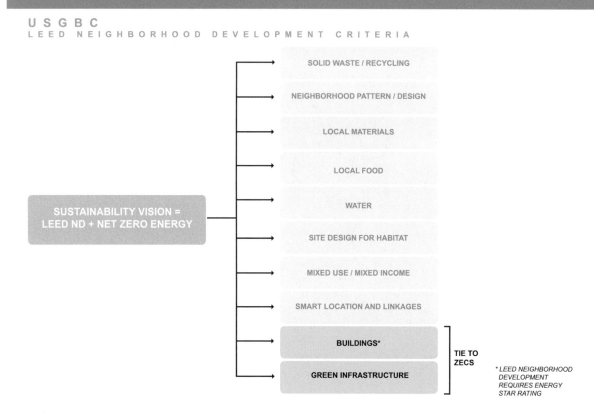

*Figure 24 - ENERGY EFFICIENCY AND SUSTAINABILITY PROGRAMS OVERLAP IN ZECS IN THE AREAS OF ENERGY, GREEN BUILDINGS, AND INFRASTRUCTURE. (Courtesy of Design Workshop and Lowry Redevelopment Authority)*

HERS or LEED certifications require energy conservation measures that are instituted during the building, design, and construction processes, and then verified with testing and certification. A ZEC developer will establish the design standards that all architects and builders will have to comply with for the project. The ratings under LEED and the application of other standards to a ZEC (Energy Star, ISI and HERS) properly and specifically provide energy conservation performance.

The ZEC planning team may use reference to these other standards and programs to describe their plan for energy conservation. In this fictitious example, the ZEC team represents the initial project design as follows:

The XYZ ZEC community will be a transit-oriented development at the endpoint of the metropolitan light-rail transportation system. The new neighborhood has adopted the classifications of LEED (Gold) Neighborhood Design, LEED (Platinum) for all municipal buildings (includes schools), and LEED (Gold) for all commercial buildings. Additionally, the new homes at XYZ will score a HERS 40 rating and be certified by the Department of Energy as Zero Energy Homes. The plans for development include many sustainability features that improve health, energy, pollution, and stress—like a 35-acre park that includes a 5-acre reflection pond, a swimming pool heated with solar thermal panels, and greenbelts to connect the community in the new 320-acre development. Every parking place will be equipped with an electric vehicle charging station.  (Illustrative language)

Conservation is the foundational requirement of a ZEC initiative. Building energy efficiency programs are familiar to builders and developers today, as evidenced by the number of projects already certified. LEED and HERS provide a means for independent, third-party verification that a building, home, or community was designed and built using strategies aimed at achieving high performance in key areas of human and environmental health: sustainable site development, water savings, energy efficiency, materials selection, and indoor environmental quality.

*Other Programs and Standards Resources*

| ORGANIZATION | WEBSITE |
| --- | --- |
| Formally known as the World Commission on Environment and Development (WCED), the Brundtland Commission's mission is to unite countries to pursue sustainable development together: | http://en.wikipedia.org/wiki/Brundtland_Com |
| USGBC (2009) LEED Neighborhood Development Rating System, Washington DC: Congress for the New Urbanism, Natural Resources Defense Council & The US Green Building Council: | http://www.usgbc.org |
| United States Department of Energy— Energy Star Program: | http://www.energy.gov |
| Passive House Institute US (PHIUS) | http://www.passivehouse.us |
| California Energy Commission—HERS– (Home Energy Rating System): | http://www.energy.ca.gov/HERS/ |
| United States Environmental Protection Agency (EPA), community water: | http://www.epa.gov/watersense |
| Institute for Sustainable Infrastructure: | http://www.sustainableinfrastructure.org/ |
| National Renewable Energy Laboratory: | http://www.nrel.gov |

*Figure 25 – OTHER PROGRAM AND STANDARDS RESOURCES: ORGANIZATIONS AND WEBSITES*

# SECTION 15
# INTERNATIONAL ZEC DEVELOPMENT

I have had a unique opportunity to travel to Europe, India and the Philippines during the past four years, and to witness—and in some cases be involved with—the development of building projects in these countries. My ideas about international development of zero energy communities have been formed during seventeen global trips, during which I spent approximately 25% of my time living abroad.

Advancement of building trades: Zero Energy Developments are complex in that they require energy features to integrate across all of the elements of buildings and infrastructure systems, rather than being something that can be bought and then bolted onto a community to make it into a ZEC. A number of disciplines converge in advancing energy efficiency in building trades, a few of which are: architecture, engineering, building construction trades (those who build with concrete, steel, wood, glass, stone, and roofing, as well as those who install the electro-mechanical equipment systems). Each trade has to adapt to energy-efficient methods, which include using special products, techniques, and adherence to precise details, plus ensure the integration of work between trades.

Regional considerations: ZECs developed in regions with distinctly different climate conditions, cultural sensitivities and economic conditions will certainly be different from their counterparts in the US.

In Europe, there is considerable evidence of green building, and that practice is growing. Driven by carbon-cap and trade accounting, and aided by a utility environment that is more of a collective than a corporate culture, Europe has deployed more renewable energy than the US has to-date.

> *In Asia and the Pacific: The international spread of sustainability is apparent at the Asian Development Bank (ADB) in Manila, Philippines. ADB is one of seven world banks, whose charter is to reduce poverty and develop sustainability in Asia and the Pacific. ADB's leaders will readily discuss their views on sustainable development and strategic plans to enhance the sustainability of their own campus, and beyond. ADB finances sustainable development at a rate of $10 billion dollars annually. (Asian Development Bank 2009)*

In India, which is developing rapidly, offices are cabins, and they are typically much smaller than the typical US offices. Residential houses are typically occupied by larger nuclear families and have different layouts than the US, and the whole gamut of electricity regulation and market pricing fluctuates. It is less certain how this will all change.

In Japan: The population is increasingly more conscientious about their energy requirements, and increasingly questions the promise of nuclear energy as a solution. A nation known for producing trend-setting technology, Japan may someday provide leadership in the technologies that underlie the development of ZECs. According to NREL, Japanese auto manufacturers have teamed with homebuilders and developers and are already selling homes and electric vehicles as a packaged product.

Developing countries: Clearly, people in developing countries could benefit from the ZEC model for their development projects. These populations face similar problems with the environment and energy and, in fact, their access to energy is more limited, the cost of energy is higher, and the power is less reliable.

Obstacles to green building: There are distinct differences in the building practices I observed while abroad, based on challenges in material and equipment availability, limitations in the skills of the building trade workers and lower standards of quality. These obstacles present many challenges to a ZEC development project that relies on precise details of execution to realize energy efficiency. There is also considerable corruption that can interfere with objective decision-making about the project procurement.

A positive attitude and teamwork can overcome many obstacles, and there are enough advocacies for green buildings occurring where at least a few are being developed despite these other problems. I believe that it will take a decade longer before developing countries value green buildings to the extent that the US, Canada and Europe do.

### Conclusion

The whole of the ZEC is dependent on each of the parts, and the sophistication of building practices overall needs to evolve further off-shore, as it has in the US and Europe. At that point, I believe we can expect to see rapid uptake of ZEC communities abroad.

I am very hopeful that the economic and environmental advantages of a ZEC will be showcased across many developments in America and that there will be effective technology transfer to facilitate a mass-adoption of ZECs abroad.

Notes

# SECTION 16
# CERTIFICATION OF A ZEC

*A certified Zero Energy Community (ZEC) may command distinct value in the real estate marketplace of the future. America's experience in creating 9 billion square feet of commercial green buildings, using the United States Green Building Council (USGBC) LEED rating system, demonstrates the proliferation of this trend and has caused a market phenomena where, according to the Urban Land Institute (2009), green features are considered essential in Class A Office buildings. Tenants are attracted to building features that reduce water and electricity use (thereby lowering utility bills that are passed through to them) and create a healthy workplace for their employees. (Brett and Schmitz 2009)*

*Green building attributes increase the market value of those real estate properties in several ways, in addition to improving the interest of prospective tenants. Many commercial lenders only finance LEED certified buildings today, and large Real Estate Investment Trusts (REITs), such as the Piedmont Office Realty Trust, are upgrading every building they own to the LEED standard, according to their senior manager Jason Williams. (Williams 2013)*

Considering a green building is much less expensive to operate than its counterpart, and that green buildings are healthier, it is not surprising that a green building is more marketable and more easily financed than an otherwise equivalent non-green building. This distinction, particularly when LEED certified by a respected third party agency using a widely accepted methodology, brings considerable value to the building owners and occupants.

*Residences are not as far along the green building adoption curve as commercial office buildings, but the inquiries that the Lowry ZEC developers in Denver, Colorado, made to seventeen prominent homebuilders during 2012 (mostly national firms), indicated that green building standards are becoming essential requirements in homebuilders' product lines. According to David Andrews, Project Manager of the Lowry Redevelopment Agency, all seventeen homebuilders proposed home products for the ZEC that would be certified at the Household Energy Rating Scores (HERS) between 40 - 49, a score that represents the absolute state-of-the-art in super-insulated homes. (Andrews 2013)*

There is time and cost associated with certification; until certification is as common for homes, as it is for office buildings, planners may pass on or postpone a certification process. There are many potential certifications, and a struggle can ensue to determine the best programs. A whole community full of zero energy buildings may be able to get several certifications for their ZEC's buildings and infrastructure (or one), covering the entire ZEC.

Certification is an investment that, if successful, will materially improve the value of property in a ZEC community for the long-term. Returning much more than it will cost, certification stands to add authenticity to the claims of the developers, sellers and lessors. I expect that such certifications will be a market requirement by year 2025, just as LEED is necessary in the office building market in 2013. Certification also allows easier cross-comparison by consumers.

The United States Green Building Council (USGBC) LEED point system, the DOE Solar Communities Program, the Institute for Sustainable Infrastructure (ISI) Envision Rating System, and many other programs measure achievement using a Bronze, Silver, Gold or Platinum scale to distinguish the level of quality—and so could a ZEC rating score.

One group that seeks to develop a ZEC certification program is TowardZero.org (TZO), a non-profit sustainability organization that mentors, consults, and publishes information about net-zero development. TZO is currently coordinating the planned certification process with a number of other agencies that provide various rating and certification systems for cities, neighborhoods, and buildings. TZO is essentially providing stewardship of the ZEC certification process. However, at the time of this guide's first edition, no agency provides a comprehensive program or process for ZEC certification.

Any group that decides to defer certification must be very diligent in their record keeping so that documentation of every step during planning phases can demonstrate compliance with ZEC certification requirements, if that certification process is to occur later. Skip Spensley, a Denver professor, environmentalist, attorney, and engineer who has provided leadership in many large developments (including Stapleton and the Denver International Airport) recommends beginning the certification process at the outset of the project and seeking certification of the ZEC plan from a respected entity before any building is undertaken.

> *Until such an entity is identified, all efforts should be made to comply with the tenets of a ZEC as explained by NREL. This reduces the risk of doing anything unnecessary or non-compliant with certification requirements. The risk of failing the criteria for certification would be much less likely with a ZEC plan that complies with NREL's definition of a ZEC already in place. Spensley suggests that putting a certified ZEC plan in place early in the development process provides good will for all actors within the project development and adds brand-value to the ZEC project. (Spensley 2013)*

At least one community developer has expressed doubts about ZEC certification, considering it an unnecessary risk. Because the processes and certifying agencies are not yet completely delineated, there is, in the developer's opinion, a real lack of certainty in the certification process. Although this developer is not pursuing a ZEC certification, they have developed design guidelines that reference LEED (ND) and LEED (Gold) commercial building and require a residential HERS rating score of 40, plus a requirement for 2,500 square feet of solar panels for each home in the community.

Because an on-grid ZEC will interconnect with the grid when completed, the ZEC's net-zero progress will be confirmed upon completion by averaging historical meter readings and measuring progress against milestones in a ZEC plan.

The City of Fort Collins, Colorado, created a zero energy district called "Fort ZED" without using any standard or certification. The leadership of Fort Zed is now contemplating an expansion to the borders of that development and has made public their belief that by not applying any zero-net energy guidelines, they have more flexibility for improvements (Dorsey 2012).

It is important for me to note that Fort ZED is called a Zero Energy District and that the initial planning occurred in 2007, before NREL's definition of a ZEC had been published. The spokespersons I conferred with were completely unaware that a ZEC definition, or any plan for a certification standard, even existed.

The measurements used to make ZEC certification should include the quantification of avoided energy use through conservation, the net-balance of renewable energy generation and total energy resource required, and the ability to provide continuous improvement through all technology and improvement of behaviors—and use of the Bronze, Silver, Gold or Platinum scale to distinguish the level of quality.

### Need and Benefits of Certification

Numerous ZEC and ZEC-like projects referenced in this guide were developed without using NREL's ZEC definition. Although those projects are commendable for leading the way to better communities both today and tomorrow, creating a pathway or certification program to develop a ZEC through a set of benchmarks and/or standards for ZECs will legitimize them and establish replicable models for ZEC development.

Certification will help to create a replicable pathway for developing ZECs and will make it easier for lending institutions, consumers and real estate professionals to understand and appraise the value and benefits of being a part of a ZEC. Certification will also help improve investment performance in ZECs, reducing risk for financiers can help to lower interest rates and provide superior terms and conditions. Furthermore, participants in these ZEC programs may celebrate their accomplishment and validate the quality of the outcome through the act of a certification.

To monetize added value to the buildings and land within the community, and to the community itself, ZECs should be developed in accordance with, and certified by, a standard rating system using meaningful criteria to measure the performance of the ZEC. Given the benefits that standardization in the ZEC process can create, it is likely that ZEC certification will become available quickly and that qualified certifiers will be available.

The certification of a ZEC includes the people, buildings, and electric vehicle-charging infrastructure, and therefore it may not be necessary to obtain certification of each building inside the ZEC when the overall result is indicated to be successful by ZEC certification.

Notes

# SECTION 17

# COMMUNICATIONS FOR A ZEC

*Communications play a vital role in creating cohesion and consistency within
a Zero Energy Community (ZEC) project plan in a manner that will bestow
a paradigm for future initiatives. Organizing the methodology and resources
required to assure effective communications is critical to creating a mission-driven
community; leadership's communications must echo a shared vision that has
brought this population to courageous commitment and action. (Lee 2006)*

ZECs require, by definition, behavioral change and employ technologies and techniques that may be new territory for ZEC constituents. The effectiveness of the ZEC rests upon shared and unilateral goals, values and requirements. Therefore, a ZEC communications team must involve ZEC participants and stakeholders in education across multiple subjects and actionable results. Effective communications programs can help a community create and maintain an ongoing conversation about ZEC development, operation and results.

Developing a ZEC requires the creation and management of a team and the development of the ZEC's relationship to the internal and external communities. Learning, collaboration, cooperation, innovation and efficiency are all expected outcomes of effective communications.

There will be a continuous need to communicate with those who create, live or work in a ZEC, as people need to implement and monitor their progress toward the zero-net goal. Energy goals are central to the ZEC, but other aspects of the community will undoubtedly need to be addressed.

**Key Communications Needs Will Fall into the Following Categories:**
(1.) Marketing and promotions: Including program development, business development and public relations activities and information about ZEC news and events.
(2.) Community development: Including information regarding the ZEC's leadership and development, information about community performance and issues affecting the occupants of the ZEC, and information about problems, progress and accomplishments of the ZEC.
(3.) Educational outreach: Including information related to the community sustainability goals, plus information about innovation, new technologies, best practices, and specific information about this and other ZECs.

The three key goals of a ZEC are energy conservation, renewable energy supply, and a transition to electric vehicles. The achievement of each of these goals requires the modification of peoples' behaviors, and therefore communications need to be informative—and motivational.

In my own experience as a leader of start-up businesses and an organizer of development projects, productive communication is the key to effective teamwork. Socializing information about project goals, requirements and concepts amongst a group, and soliciting inputs, helps people to feel they understand, and are a part of, that vision—even when that vision and definition are emerging. When people feel included, they tend to assume ownership of the goals; they become more involved, more committed, and they are more likely to commit and develop a true sense of teamwork.

*Peter Senge's book and workshop guide, The Fifth Discipline: The Art and Practice of the Learning Organization, explains how teams can go beyond their usual mental models about roles and possibilities and develop meaningful dialogue, where new understandings about systems and solutions emerge, and the teams then aspire to a shared vision and innovation. Communications and organizational development involve such a strong linkage, which leaders must address together. (Senge 2009)*

Given the importance of communications, ZEC planners should focus considerable effort on developing and structuring a plan for the management of communications before communications problems begin to interfere with their progress.

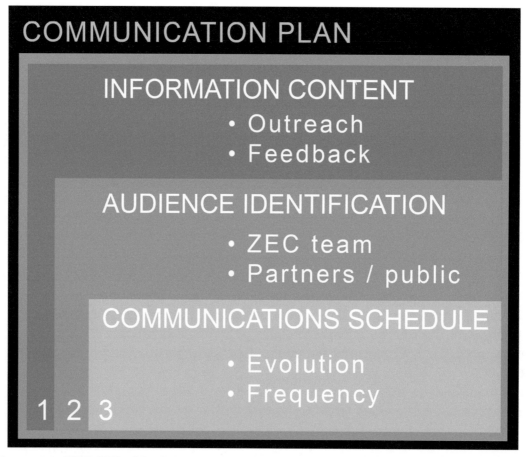

*Figure 26 -THE KEY CONSIDERATIONS OF A ZEC COMMUNICATIONS PLAN*

### Communications Plans

Communications plans provide a structure for organizational development, combining the information about audiences, messaging content and timing in a cohesive structure. To be effective, they must involve the following elements:

Audience identification: Who is involved in the ZEC, and who else will need to know about it? How are those audiences grouped? ZEC initiators provide the key communications at the outset of a ZEC, but as the initiative is developed, the volume of information is increased, and the number of distinct audiences may grow. The core ZEC team includes leadership, management, operations, business, legal, energy and technical experts, and may include homeowner's associations, investors, and advertising and public relations agencies, architects, engineers, builders, technical advisors, suppliers, vendors, volunteers and local media.

Other stakeholders in a ZEC may include residents, merchants, vendors, contractors, consultants and project managers. Additionally, agencies addressing urban planning, public works, economic development, building permits, fire inspections, transportation and sustainability may become involved.

Information content development: What information will need to be shared within the ZEC team and externally?  What content needs to be created about the ZEC team's vision, mission and values, and the benefits and features that underlie the value proposition for the ZEC?  What is the description of the plan for the site, the buildings, and the energy technologies? What programs and services are envisioned for the community? Is there information about prices, policies, promotions? What are the public relations, educational and other outreach goals? Define all content required and determine who will develop it.

Communications schedule: Many of the ZEC communications can occur according to a routine schedule, with daily, weekly, monthly, quarterly or annual frequency. As the ZEC advances from a nascent stage to an established institution, the needs will change. A schedule provides some discipline to the communications process and helps to keep communications fresh and meaningful, and it eases the workload for those involved in creating, disseminating and responding to communications. The communication plan may also include a message matrix that relates the key types of messages to the specific project sub-teams and stakeholder groups, so that each group communicates often enough about subjects that are most relevant to them, with more general communication occurring among multiple or all stakeholder groups.

Meeting planning: This is another very important aspect of the communications plan. Assuring productive meetings requires forethought in preparation, meeting conduct and post-meeting action. In addition to planning logistics (such as conference-room reservations, audio-video equipment and refreshments), a ZEC team can assure that participants understand meeting objectives, resolution and action requirements, by preparing agendas, providing materials in advance, and keeping meeting minutes. An intentional plan to improve meetings can create great value.

Creating conversation: What type of feedback could be valuable and how will it be incorporated?  How can communications be bi-directional? Intentionally soliciting feedback and developing dependable processes to review and respond to that information could provide valuable insight to the ZEC team.

| SUBJECT | AUDIENCE | INFORMATION CONTENT | FREQUENCY |
|---------|----------|---------------------|-----------|
| Executive report of planning and operations, and finance | ZEC Leadership Team (confidential) | Submission of report and discussion at a meeting | Monthly, on the 2nd Tuesday |
| General Development Plan (GDP) | Entire community | Structured public process including posting of GDP and public hearings | One-time approval required prior to construction activities |
| Inauguration of the ZEC | All Stakeholders | Reception with line-up of speakers and presentation of television advertising by the homebuilders. | One-time event synchronized with the opening of the Visitor Center. |

*Figure 27 - SIGNIFICANT AMOUNTS OF INFORMATION CONTENT, MANY USERS, INFORMATION CONFIDENTIALITY CONCERNS, AND TIMING OF COMMUNICATIONS CAN BE MANAGED WITH A COMMUNICATIONS PLAN.*

Record keeping: It is a good idea for any business, especially one involving a public interest, to keep accurate records of all communications. Organization of the message data and regular archiving should be a part of the administrative plan.

Information security: A ZEC's communications program must balance organizational transparency and information security; some information will be confidential so as not to adversely affect interests, while other information can bring great value when disseminated widely. The "company confidential" classification of confidential information is intended to label documents that can only be shared with internal audiences.

Digital communications: There has been no time previously when communications were as instantaneous, personalized, interactive, and effective, as they are today. The means of communications are vast, and ever expanding. Communication can be accomplished via mail, email, websites, blogs, post updates or articles, tweet, vote, search, plus ZECs can transact and/or share all of the information concerning the ZEC. Additionally, ZECs can create meeting invitations, web-conferences, Google Hangouts and many more options.

Social media: The US is almost completely Internet-enabled today, and this provides connections for people, information and groups. The range of uses for social media has quickly grown; now there are many new ways to communicate, all of which can be used together, powered by incredible automation. The choice of communications media is ever-expanding, and those responsible for managing the communication could become overwhelmed with the decisions about what social media platforms to use. Social media operates as a communication tool—and much more.

Joshua Pollock, Goddard College graduate student, social media expert, and student of environmental studies, developed a method for establishing ZEC management and communications using social media as a meta-structure for creating projects, project management, and engaging people in those sustainability projects.

Pollock uses social media as a project management tool, where the steps of the project, the assignment of responsibilities, and schedule are all incorporated. His vision of re-localizing projects across multiple cities is very compelling, and the benefits include sharing knowledge, connections and skill-sets, and continually developing a database of information. He makes a strong case for social media becoming integral to the program of developing a ZEC:

Social media capabilities have the potential to allow the local ZEC teams to draw from the experiences of teams in other locations, and bring all stakeholders into the project, instead of simply informing them, as is often the case with social media. In this way, a social media platform allows people to contribute expertise, connections or funding.

As more projects use social media, the greater the opportunity to use the analytic information—that social networks usually gather for the purpose of targeting ads—for connecting people working towards common goals. This untapped resource could be used to predict what types of people are needed to successfully implement a project and make artificially intelligent suggestions for who could fill those roles by drawing from those already connected to the project via existing social networks.

Social media meta-structure that tracks the work of the ZEC should be aimed at turning supporters into contributors and sharing experience between ZEC teams in different sites. For example, a team contemplating using ground-mount solar panels learns about the maintenance problems experienced by a ZEC in another city—and how another overcomes the issue by using solar carports.

> *Sharing this type of information via a truly social website will allow the developing Zero-Energy Communities to develop a dynamic living document that evolves, based on the experience of those using it. A non-hierarchical organization of information allows searching based on criteria—which may include geography, type of ZEC project, size, functions and progress points—and which would allow the document to adapt itself, based on the needs of whomever is currently reading it. (Pollock 2013)*

The thought of simultaneously integrating and maintaining interfaces with Facebook, LinkedIn, Google+, Plaxo, Maven, and Twitter, may overwhelm the average person managing communications, but automation tools have made it much easier to manage this type of challenge. There are many different tools that automatically interface all social media platforms together, allowing users to self-select the type of information they require, the frequency of notification, and the social media platforms that work best for them.

Omni-directional communications: Because social media runs upstream and downstream, it breaks the mold of broadcast style communications. Administrators and moderators determine exactly what can be controlled by each person in the network, so communications are still reasonably manageable.

Websites: Website development tools, including open-architecture programs such as WordPress, can be created and updated easily using standard word-processing programs. The tool supports website designs and blogging, and there is a large choice of pre-designed website templates, customizable layouts and graphic designs, and various styles of buttons, information channels and many other features.

Plug-ins: Quick and easy to use applications are available to enable new functionality in your website and social media program—all automatically. Among the tools designed for this purpose is BuddyPress, which allows users to sign up and create profiles with their communications preferences. They can post messages, make connections, create interactive groups and easily build a community for their ZEC. Many types of plug-ins are available. (See plug-ins under www. wordpress.com and wordpress.org)

Crowdsourcing: Using crowdsourcing plug-ins with social media allows one to locate and contract resources of a specialist to assist with any ZEC task. For example, if you needed a graphic illustration designed, one could use crowdsourcing to find an artist, negotiate a price and arrange digital delivery.

Crowdfunding: This technique allows a large number of people to invest a small amount of money. This methodology might be used to provide the funding to pay for a ZEC feasibility study or for an existing ZEC to pay for a special program, such as the installation of solar thermal water heating to warm the community swimming pool.

> *Custom plug-ins: Programmers who develop open source plug-ins for WordPress can usually be hired to change or adapt existing apps to meet a specific need. If the original developer is unavailable, a different developer can be hired since the code is available for anyone to work with. Josh Pollock spoke about WordPress' open-architecture ecosystem: Matt Mullenweg, one of the originators and a lead developer of the WordPress software and CEO of Automattic Inc., WordPress. com's parent company, said, There are more than ten-thousand people making a living using WordPress in the world today. (Pollock 2013)*

Aggregators: Also known as news aggregators, feed aggregators, or reader aggregators, these are web search websites that can search designated sites, or keywords, or any user-defined subject matter. They can automatically and routinely search the Internet and syndicate the search results in accordance with each user's requirements and preferences. Internet search results can include links to documents, news articles, journals, blogs, podcasts, and YouTube-type videos. (An example of an aggregator site is www.alltop.com.)

Social circles: Viewing team structure according to social circles is helpful in organizational development. Using a diagram formed by concentric circles, leaders are listed in the center circle, and each successive circle provides a band for other social circles. In this diagram, the circles are labeled: Leaders, Occupants, Advisors, Partnerships, and Public groups. Define your own social circles using this technique; circles can be developed around many other special interests and project groups.

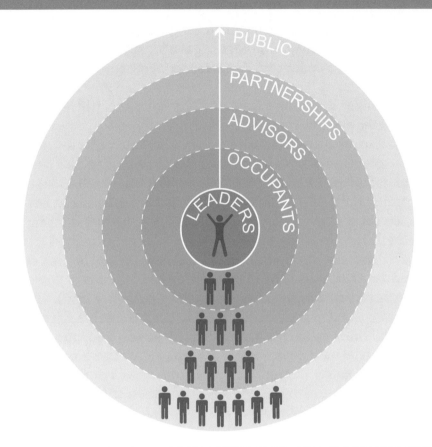

*Figure 28 - SOCIAL CIRCLES INDICATE INTERNAL AND EXTERNAL RELATIONSHIPS FOR A ZEC. CATEGORIZING THE SOCIAL CIRCLES AND MEMBERS USING SEARCHABLE "TAGS" WILL HELP ZEC TEAMS TO MANAGE COMMUNICATION PREFERENCES, CONTACT DATA, AND OTHER RELATIONAL DATA.*

Segmenting audience groups can help to coordinate communications. The leaders in the center receive more information than audiences that are more peripheral. "Social circles" refers to any special interest group formed by the sharing of common interest (usually among social media groups) through the Internet. There is a universe of relationships formed around a ZEC, and using the concentric social circles (as above), ZEC planners can organize their communications. Once categorized, a ZEC team can begin to set their communication goals for each audience.

Participant database: ZEC audiences will undoubtedly change and restructure through the evolving life of the ZEC. Given the continuity of a ZEC, it is critical to develop a proactive method to change and update the data about audiences and individuals within the communications infrastructure. The database information needs to be updated whenever people join, leave or change their relationships.

Tags: Tagging allows one to categorize, by attaching labels to each individual or organizational relationship, document, article or post. ZEC planners can more effectively search, locate and filter social media by using tags, which add significant flexibility in managing the deployment and customization of communications.

On-boarding: Processes that help new-hires and others acclimate to new work processes and workplaces are called "on-boarding." On-boarding is a pro-active orientation and training program for new employees, new contractors, business partners or a ZEC volunteer team. Incorporating people, or so-called "human capital," into a team and project requires sharing information and ideas. Making sure people have the information that will help them is a reliable way to assure success on the projects.

In my experience, on-boarding processes have proven particularly effective, and I believe them to be of great benefit to people, and to project success. If on-boarding sounds like something that only a large organization would need to use, I suggest that you test that idea and examine the impact of on-boarding with a small, formative group. In my experience, it has been crucial to brief the team members about over-arching issues of the project in order for them (and you) to understand their best value for the organization.

The "digital divide" refers to an economic inequality between groups, broadly construed in terms of access to, and use of, or knowledge of information and communication technologies (ICT) (National Telecommunications and Information Administration 1995). The separation of computer users and non-computer users exists in the US, but that gap has closed significantly. There are still people who either do not use or have limited use of a computer, and ZEC communications teams should meet people where they are, even if that is on a printed page.

*Guy Kawasaki, a founder of Apple Computers, discussed social media when he visited Denver, Colorado, in 2011. He provided a demonstration of how he uses news aggregators to locate articles of interest on the Internet. His technique for daily posting of relevant articles to his social media site is on a tight schedule— every morning. Guy reminds us all to enchant our audiences. (Kawasaki 2011)*

### Conclusion

A communicative ZEC team can facilitate dialogue, develop community support, and unify a talented and passionate team of people required to develop and operate—and continuously improve—a ZEC. Communications are central to forming a shared understanding of the ZEC among all stakeholders and to consistently motivate all toward the zero goal. An abundance of tools and resources are available, most for free, to enable effective communications that can inspire and engage all who are involved in the zero energy community.

*Questions for ZEC Communications Planners*

(1.) Who will manage communications? Who will support them in that effort? Who will manage, administer, edit, moderate, write, archive, and manage documents?

(2.) How will the resources of the ZEC core team be supplemented by professional agencies and the in-house teams of key vendors and partners?

(3.) What tools will be deployed to manage correspondence and publishing?

(4.) What is the plan for the website?

(5.) How will effective meetings be organized?

(6.) How can web-conferencing / conference calling be supported?

(7.) How will the ZEC incorporate social media?

(8.) Will crowdsourcing be deployed?

(9.) Will the website provide credit card payment and/or electronic banking?

(10.) Will crowdfunding be deployed?

(11.) Will there be a visitor center?

(12.) Will there be a newsletter?

(13.) Will there be sales brochures?

# INITIATING A ZEC PROJECT

## *Step 1 of 10*

With the knowledge you have gained from earlier sections of this guide, you should now have an understanding about all of the factors that are relevant to a ZEC. Aspects of ZEC planning that have already been covered have included energy utilization, the electric grid, energy policy, energy conservation, renewable energy, energy transfer and storage, the ZEC concept, opposition to ZECs, living in a ZEC, hybridization, electric vehicles, control systems, sustainability programs, international ZECs and financial considerations of a ZEC.

It would be delightful to know that your interest and understanding have swelled to the point that you can now imagine organizing a ZEC project. In the remaining sections, I will share a simplified ten-step process that will guide you through developing and operating a new Zero Energy Community.

The main planning requirements for a ZEC are establishing the perimeter border, developing a conservation plan, making a renewable energy plan, and creating a plan to accommodate electric vehicles. These four goals need to be forged within a cohesive plan that will respond to all needs of the community. An iterative and emergent process can help to define the ZEC, with all the key factors tested against the information that becomes available through the course of the ten-step planning process.

It is human nature to want to jump to the conclusion of such an assessment and planning process and to answer, what will the ZEC cost? How will it operate? What are the plans for building energy efficiency, renewable energy, and deployment of home control systems and other smart building technology? What would it be like to live or work there? How will the resale price of green buildings compare with conventional buildings?

First, however, an understanding of the ten-step process is important. The individual steps are subsequently outlined in this first step, Initiating a ZEC Project.

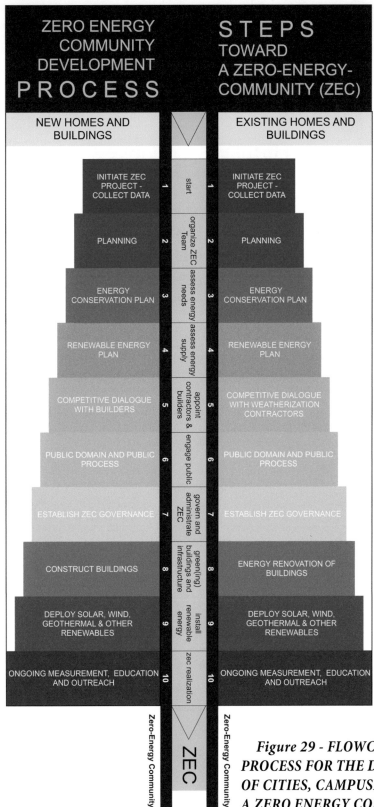

*Figure 29 - FLOWCHART FOR THE TEN-STEP
PROCESS FOR THE DEVELOPMENT AND RENEWAL
OF CITIES, CAMPUSES, AND NEIGHBORHOODS TO
A ZERO ENERGY COMMUNITY STATUS.*

Each step in the process of developing a ZEC needs to create value and reduce risk. Each step involves increasing commitments and magnitude of opportunity and risk, all of which requires management.

# 10

## *The Ten Steps of the Process*

| Step 1 | – | Initiating a ZEC Project |
|--------|---|--------------------------|
| Step 2 | – | Plan the ZEC Project |
| Step 3 | – | Energy Conservation Plan |
| Step 4 | – | Renewable Energy Plan |
| Step 5 | – | Competitive Dialogue Process |
| Step 6 | – | Public Domain \| Public Process |
| Step 7 | – | Establish ZEC Governance |
| Step 8 | – | Renovation and Construction |
| Step 9 | – | Deploy Renewable Energy |
| Step 10 | – | Measurement, Education and Outreach Communications |

**Results of the Process**

Each step in the ZEC planning and implementation process should achieve several results:

☐ Greater resolution about the ZEC's details and big picture
☐ Greater shared vision, innovation and understanding of the mission
☐ Growth of the ZEC planning and implementation team
☐ Further outreach within the immediate and affected community

## Notes

### Initiate the ZEC – Collect Data

(1.) Develop leadership and governance

(2.) Discuss the values the ZEC team and project will embrace

(3.) Formation of  ZEC initial planning team

(4.) Identify project sponsors

(5.) Identify funding sources

(6.) Identify and engage consultants

(7.) Form teams for key tracks of project

(8.) Collect and analyze data for each track of planning:
- Define goals, assess limitations, and develop list of questions
- Programming workshop
- Visualize the ZEC project
- Initial assessment of ZEC project feasibility
- Deliverables: Programming Document
- Meetings: Leadership, team development, programming workshop
- Other activities: Site visits to local green developments, similar project types
- Outcomes: Team, ZEC Vision and Program, and initial feasibility analysis

(9.) Duration: 45-60 days

(10.) See other sections: Living and Working in a ZEC (Section 2), Policy (Section 8), Exploring Renewable Energy (Section 9), Other Programs and Standards (Section 14)

### New Project Evaluation

As early as possible, ZEC planners want to secure an understanding about the vision and facts surrounding a ZEC project. What are the potential outcomes of the ZEC project?  What would be the successes?  What are the risks, and how should those risks be managed?  As the project develops through the steps, the project needs to be understood in increasing concrete details; detail is required to price, build and operate it. During the initial steps, the team should be satisfied with many unanswered questions, as the details will unfold during each successive step forward.

Notes

### Key Questions for the ZEC Team

During Step 1, the ZEC Team should address the following important questions:

(1.) What are the needs?

(2.) Who is going to be the team leader for each of the six tracks (regulation, marketing, project management, financial, technology, and operations), and who can support those team leaders best?

(3.) What local organizations and agencies are synergistic with the ZEC?

(4.) What consultants are required?

(5.) Is the ZEC a renovation, a new development, or a combination of both?

(6.) What are the proposals for establishing the border to define the ZEC?

(7.) What will be the governance and authority for energy purchase, contracts and compliance with building standards?

(8.) What are the peer projects to be visited? (green buildings, ZECs, solar gardens)

(9.) What is the electricity structure desired of the ZEC and:

    A. What are the basic regulatory parameters?

    B. What configuration of the ZEC electrical system is necessary or desirable?

    C. On-grid using net metering or virtual net metering?

    D. On-grid and off-grid with a stand-alone energy system and net metering?

    E. Off-grid with a stand-alone energy system?

### Visualizing a ZEC Project

The possibilities, opportunities, features, benefits and values of a ZEC need to be shared and comprehensible in order to create, design, fund, approve and implement the ZEC project vision. A high-level understanding of a ZEC's features and benefits emerges through understanding six tracks: regulation, marketing, project management, financial, technology, and operations, which should cover comprehensively every aspect. Throughout the course of the project, the team should create plans, diagrams, charts and other presentation materials that help others visualize the options and solutions for the ZEC.

#### Project Room

Set up an area dedicated to the ZEC project. Display ZEC planning materials on the walls as they are developed, forming a visual representation of the ZEC's evolution.

#### Planning Tracks

Creating a small number of planning tracks for the ZEC project makes an otherwise complex requirement easier to explore and understand. Planning tracks create a set of optics with which the ZEC can be understood early in the process. The six planning tracks provide an organized approach to organizing and implementing a ZEC project that can be used in every step of the development. The planning teams initially focus on the individual tracks, determining the facts, needs, goals, problems and solution concepts associated with each track.

The project takes greater form when the teams working on each track discuss their ideas with teams working on other tracks. By crosscutting the correlation of each track against the other five, the planners can gain a greater understanding of the full possibilities and challenges for the ZEC project.

Ask planning teams to develop documentation of their questions, findings and recommendations every time they meet. It is helpful when the planning tracks adopt a uniform report style so the documents can be more easily compared and combined.

*Figure 30 - INTER-RELATED EVALUATION TRACKS FOR A ZEC PROJECT – SIX PLANNING TRACKS*

This diagram indicates the tracks that correspond to the key aspects of a place-making project like a ZEC. Each planning track team should analyze the situation regarding the ZEC, determining the known and unknown information that will be required to define and develop the project.

The planning tracks usually correlate with the respective skills and experience of individuals and consultant firms assigned to the sub-groups include the following as a recommended minimum:

### Regulation
- ☐ Local regulation
- ☐ District formation (tax or energy)
- ☐ Power purchase

### Marketing
- ☐ Market evaluation - strategy
- ☐ ZEC product brand

### Project Management
- ☐ Definition of project scope
- ☐ Resource management
- ☐ Scheduling
- ☐ Budget management

### Financial
- ☐ Economic modeling
- ☐ Procurement process
- ☐ Managing tax credits, rebates and grants

### Technology
- ☐ Energy efficiency
- ☐ Renewable energy
- ☐ Control systems

### Operations
- ☐ Cost and performance of operations
- ☐ Services and service quality levels

## Notes

### The ZEC Project Team

Organizing the team: A cadre of people who are the initiators must determine what the requirements of the project include and whom they can attract as a paid or volunteer team member.

### Who Is Needed?

Intelligent people with complementary, cross-disciplinary skills enhance the functionality of a ZEC team, especially when those individuals have a stake in the outcome. In addition to the core team, it is likely that several other people will be imperative to the project development process. These may include, but are not limited to, leaders and experts such as a civic leader, a project facilitator, a real estate development expert, commercial and/or residential architects, landscape planners, and sustainability experts with experience in LEED ND and building rating and certification. The people involved with a ZEC depends upon many variables about the scope of the project.

### Creating a ZEC Team

Use the cadre of initiators and sponsors to set up functional accountability and reporting structures. Organize the project team to correlate with planning tracks to enhance and accelerate the project planning and implementation and produce a more comprehensive and satisfactory result. Below is a conceptual design for the ZEC development team.

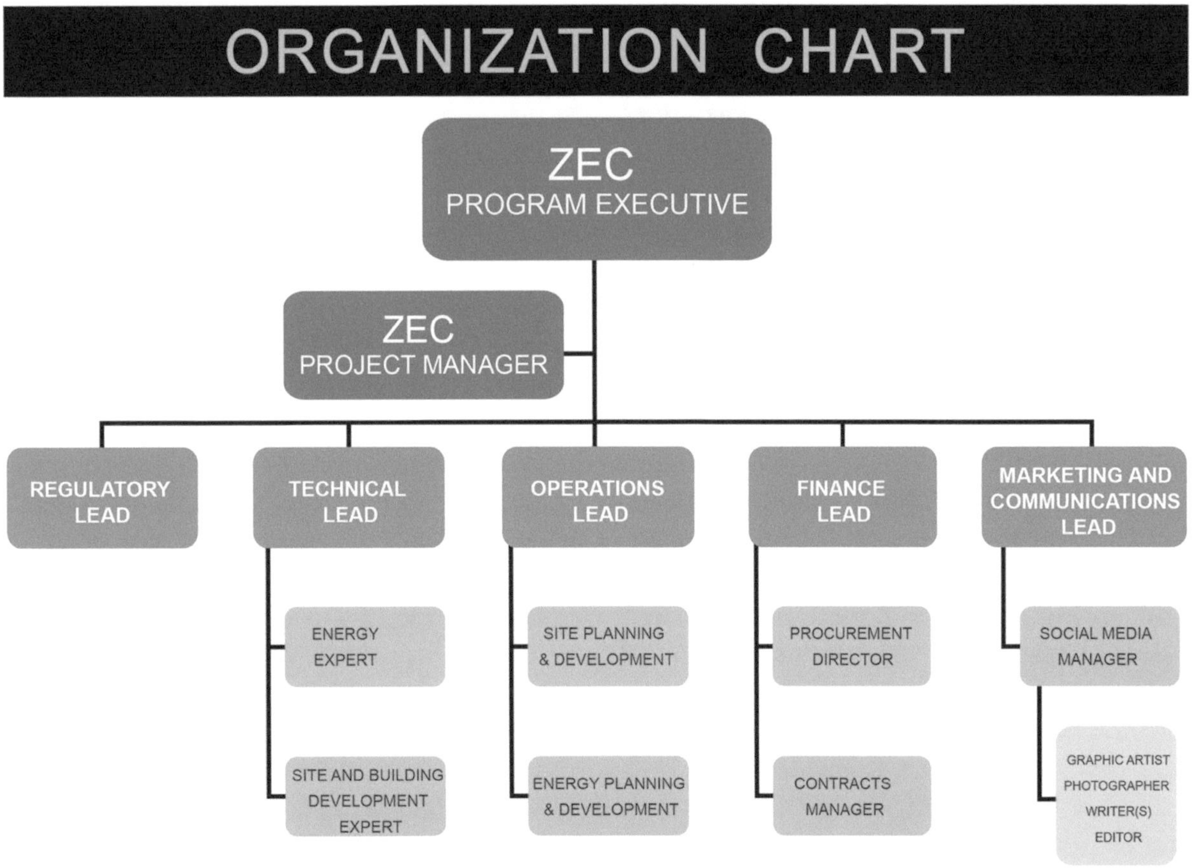

## ORGANIZATION CHART

*Figure 31 - ILLUSTRATIVE FUNCTIONAL ORGANIZATION CHART FOR A ZEC DEVELOPMENT PLANNING TEAM*

### *The Functional Organization*

The functional organization defines roles that may be shared among more than one person and may only require a fraction of each team member's time to perform the requirements. The requirements will vary, based on the size, situation, and complexity of the ZEC undertaking. For example, a plan to retrofit a ZEC for a small village or corporate campus would require less expertise and effort than that required to retrofit the entire District of Columbia. Likewise, a new Greenfield development of 5,000 homes and mixed-use buildings would require far less specialized development expertise than an existing neighborhood or medical center project.

Regardless of the size of the project, it is recommended that ZEC planners engage the support of project sponsors who can provide leadership and influence. The support of the mayor's office, for instance, would be very valuable.

Appointing a Program Executive(s) who would provide leadership and over-arching management responsibility is recommended for all ZEC projects. The person(s) serving in this role may also be organized as a board, which would accommodate several executives, or the Program Executive(s) could report to a board.

A capable project manager is required to plan and communicate the ZEC project activities. A legal expert who can evaluate, plan and manage the regulatory and legal aspects of the ZEC project is necessary because the laws related to energy, real property and development are complex and constantly changing. The team should also have a technical leader who is capable of managing technical requirements related to design and construction of buildings and renewable energy. In the case of a large project, an operations leader overseeing the site and energy planning may be required. A finance leader capable of managing procurement and contract-related matters, and a marketing and communications leader round out the management tier for the project. Support teams will assist these managers, as illustrated in the previous functional organization chart.

ZEC planners will want to make an organizational chart that is the best fit for their particular circumstance. That organization can respond to the exact nature of the ZEC, the available resources, and skills available from various staffers, workers, consultants, and volunteers. Planning a ZEC requies a cross-disciplinary team.

Depending on the situation, a new business unit, department initiative, or an altogether new entity must be formed to provide leadership, financial authority, ownership, and operating infrastructure for the project. The entity may be a company, non-profit, or an adaptation of an existing association.

At any point in a ZEC project, information or events could arise that could cause the ZEC project to become shelved, temporarily or indefinitely. Feasibility analysis efforts can quickly root out potential fatal flaws of the project, but market changes cannot always be anticipated. For instance, during 2009, I was involved with a large sustainable mixed-use project that included eight-thousand new homes. Suddenly, the US housing market crashed, making that project unfeasible. Commencing the feasibility analysis early, and keeping it in play for as long as possible, yields an early indication of feasibility and a sustained process to continue providing due-diligence analysis of the ZEC project.

### Technical analysis

The process of design and engineering analysis can become extremely informative. The documentation of program requirements tends to ground the technical options for comparison. Design solutions for the ZEC program may be related to land use plans, building design concepts, programs for education about energy conservation, or use case studies for smart buildings and control systems.

### Workshops

The best way to produce planning results rapidly is to hold workshops. Bringing sponsors, advisors, and stakeholders into a workshop environment usually brings people and their ideas together. The use of workshops helps teams manage uncertainties because there is more likely to be a consensus and the tight schedules for reporting out daily results in a workshop can drive teams to make assumptions and decisions.

### Inclusive process

There may be the need for certain command decisions in order to develop a ZEC. I urge the team to challenge these decisions if there is a fatal flaw in them, but otherwise, to partake in an inclusive process where all participants use their unique rationale, intuition, and intelligence to inform better processes and to drive great decisions.

### Other Planning Processes

Future capabilities: As the ZEC project advances, new questions and ideas about project features will emerge; some features will be debatable, not fundable, and/or not practical to implement initially. When there is a desire to reserve the option to add a feature to the ZEC at a later stage, such feature(s) can be designated as "future" in the program document. As the project gains momentum, possibilities may expand and the ZEC can be planned to meet current needs, while also planning how the ZEC will support future ones.

Notes

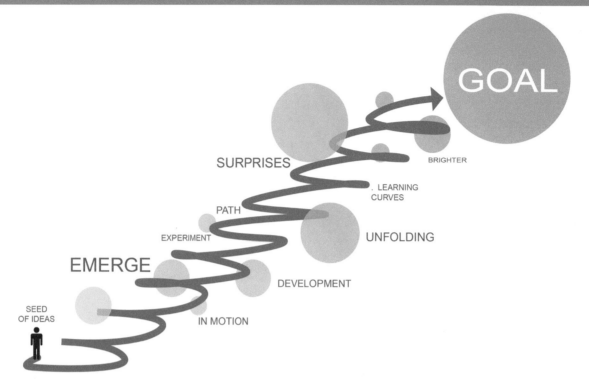

*Figure 32 - THE ZEC TEAM CIRCLES AROUND THEIR GOALS AND CHALLENGES AND A ZEC PLAN EMERGES*

### An Iterative Process

The early stages of planning can seem like a chicken-and-egg process. To make progress, a planning team must push forward and address the elements of the project as best they can by making assumptions and  discovering the issues and their inter-relationships. It takes putting a stake in the sand and creating a concept. The program of requirements evolves out of that—usually in several cycles.

### Summary

The implementation of the Communications Plan described in the previous section is extremely important during all of the remaining steps of implementing the ZEC.

Before proceeding with Step 1, the initiator(s) must organize a means to guide the process, engage sponsors, secure financing, or purchase property for the ZEC. The leadership team may have to operate without funding initially in order to secure funding to support the ten-step ZEC planning study.

It takes courage to initiate and collect data about a ZEC that may be just at the stage of an idea. The advocacy of a ZEC project prior to the study of cost and benefits, technical and regulatory parameters, requires a leap of faith by the initiators. The feasibility may not be understood or ensured at the outset of the ZEC project.

The ZEC project may commence when the first initiator communicates the idea to others. Initiating a project before a budget is appropriated is an act of leadership.

Now it is time to proceed to Step 2.

# PLAN THE ZEC PROJECT

## Step 2 of 10

The second step of the ten steps begins with a conception of the project. The conception should result in a project document from the ZEC team's initial attempt to examine the possibilities, goals, and vision for the ZEC project during Step 1. Step 2 in the ten-step process organizes the planning information, develops the ideas for the design into more exacting parameters, and solidifies the design concepts and other project information in a manner that is easier to share and understand. The second step also begins to define each ZEC team's individual project plan, describing the schedule and budget for future planning and implementation activities, and project deliverables.

### Step 2: Objectives: Planning the ZEC

(1.) Incorporate new team members on track-planning teams.

(2.) Socialize the information in the Program Document by discussing, questioning and creating a framework for shared agreement about it.

(3.) Understand options for the ZEC layout.

(4.) Assure that the ZEC plan responds to requirements for each planning track.

(5.) Understand the ZEC's key design features and benefits.

(6.) Understand the additional design work required.

(7.) Understand the project timeline.

(8.) Develop a "Design Development Document" that responds to the requirements in the Program Document, with new information about the specific planning solutions.

(9.) Determine the ZEC border.

(10.) Prepare tables with information regarding the ZEC site (buildings, area, building volumes, roof areas and parking spaces).

(11.) Deliverables: Design Development Document.

(12.) Meetings: Leadership, team development, planning-track teams, design workshops, and group meetings.

(13.) Other activities: Team building, obtain maps, desk review, and research.

(14.) Outcomes: Team development and Design Development Document.

(15.) Duration: 45-60 days.

(16.) See other sections: About the ZEC Concept (Section 1), Financial Considerations – Zeconomics (Section 3), Conservation (Section 6), Electric Vehicles (Section 10)

### Key Questions and Issues for the ZEC Team

(1.)  What people and skill sets should be added to the team during Step 2?

(2.)  What other local organizations, agencies, firms and people are synergistic with the ZEC?

(3.)  What additional consultants are required?

(4.)  How can the ZEC initiative strengthen connections with the community?

(5.)  What are the gaps and inconsistencies in the Program Document?

(6.)  Who are the best leaders for each of the three design teams?

(7.)  How can each planning team incorporate members from track teams?

(8.)  How can the ideas coming from the three planning teams be firewalled?

(9.)  What are the unified requirements of each of the planning teams?

(10.)  Is the ZEC a renovation, a new development or a combination of both?

(11.)  What is the proposal for the borders of the ZEC?

(12.)  How can the ZEC ensure acceptable energy costs for all ZEC occupants?

(13.)  How can the ZEC ensure acceptable weatherization or construction costs for all ZEC occupants?

(14.)  How can the ZEC plan address low-income occupants?

(15.)  Activities for planning the ZEC.

(16.)  Resolve concerns and add content to update the Design Development Document.

(17.)  As a group, discuss status and goals for each track team.

(18.)  Presentation on the ZEC plan responds to each track of the Program Document.

(19.)  Description of the ZEC's key design features.

(20.)  Description of the ZEC's benefits.

(21.)  List of additional design work required.

(22.)  Discuss the project timeline.

(23.)  Have track teams work individually to complete their updates to planning.

(24.)  As a group, review and approve the updates made by each track team.

(25.)  Temporarily dissolve the track teams and appoint three design team leaders.

(26.)  Split the six track teams up among three new design teams, and others.

(27.)  Have each of the three design teams create a plan for the ZEC.

(28.)  Have design groups present and answer questions about the three alternative ZEC plans.

(29.)  Combine the best information in the three schemes into one design scheme with options.

(30.)  Choose design elements from the three designs and combine into one new plan.

### ZEC Planning Deliverables

The deliverables for Step 2 include information that further resolves the goals of the ZEC project in a broad context, while responding to the emerging requirements along each of the six planning tracks. The deliverable should reflect creativity in the approach to the site plan, conservation, renewable energy and the accommodation of electric vehicles. A number of possibilities for further development should be identified so they can be explored in the next steps of the project.

### Step 2: Deliverables Include:

(1.)  An update of the Program Document that clearly identifies the project goals.

(2.)  Organize along six planning tracks.

(3.)  Incorporate program goals that emerge through the design process.

(4.)  Identify the measures and metrics that can be applied to each goal.

(5.)  Identify options to be explored.

(6.)  List questions that will need to be answered in subsequent steps.

(7.)  A Design Development Document that consists of:

    A. One or more site plans showing ZEC borders, buildings, streets, parks, paths and greenways (planned and existing).

    B. Diagrams identifying energy conservation concepts and energy equipment to be considered for all residential and commercial buildings.

    C. Identification of large onsite areas where solar panel arrays could be installed to supplement those mounted on buildings.

    D. Identification of areas where onsite wind turbines could be installed.

    E. Identification of all parking places requiring electrification for vehicle charging.

    F. Identification of additional sites where the ZEC might develop offsite renewable energy.

    G. List of options to be considered, including items such as:

        i.  Solar carports

        ii.  Solar street lighting

        iii. "Purple-pipe" water management

        iv. Hydro and micro-hydro-electric facilities

        v.  District ground loops

        vi. Battery vault locations

        vii. Generator locations

(8.)  Make tables that provide the size of the following:

    A. Size of the ZEC overall

    B. Building footprint area by type

    C. Estimated internal volume for all buildings by type

    D. Building rooftop area in square feet by type

    E. Number of private and public parking spaces by category:

        i.  Garage

        ii.  Solar carport

        iii. Uncovered parking

(9.)   Develop an outline of the energy conservation plan, including education and outreach for every type of land use.  Identify goals, training programs, incentives, monitoring and reporting appropriate to land uses such as:
   A.  Residential (attached/non-attached, apartment, condo)
   B.  School
   C.  Office
   D.  Retail
   E.  Park

(10.)  Write an Executive Summary addressing:
   A.  The ZEC project goals for conservation of energy, the use of renewable energy, and the accommodation of electric vehicles.
   B.  A profile of the ZEC leadership and team assigned to the project.
   C.  An estimate of the known cost related to:
      i.  Planning the ZEC
      ii.  Implementing the ZEC
      iii.  Financing community projects
      iv.  Financial implications to renters, homeowners and commercial occupants

(11.) Draft and continuously update the Design Development Document should be throughout the process of Step 2. This document will describe the information known about the project requirements, options, and remaining information requirements.

(12.)  Add sections of information to the document in each of the remaining eight future steps. The document should be organized with this further development in mind.

(13.)  Organize and include various maps, plans, sketches and photos to further explain the design intent. The large format plans should be presented along with information written in the Design Development Document.

(14.)  The Design Development documents and drawings are usually expressions of intention, illustrations of planning studies, and tables of information. Design Development documents and drawings usually do not describe how buildings, systems, or landscape features are actually built. The focus of this work is defining the performance attributes and creative ideas for achieving that performance, and measurably good results.

Notes

### *Design Development Team*

After determining the basic design parameters, dissolve the three planning teams and create a new design development team. Form a new design development team that includes the three design team leaders from the previous exercise and one or more persons from each of the six planning-track teams. This team of nine or more people will be responsible for developing the design concept and program goals to a greater level of detail. They will respond to the discussions made during the design review.

During this phase, the planning-track teams and others will be reassembled as design teams. Three design teams usually produce a range of creative and practical results that are hard to achieve when only one or two options are being pursued. The three teams should have a fairly equal diversity and depth of talent. Ideally, each design team will include at least one person from each of the six track teams. Planning professionals refer to this type of workshop as a "charrette," a French term that translates to "cart," and refers to the collaboration among art students that occurs as they walk together with the cart that hauls the students' art projects to the school's storage barn each evening.

### *More about the design teams:*

☐ Invite community leaders, architects, planners and community members to participate, as it will be very valuable.

☐ Design teams usually operate in a workshop style, meeting for one or more days in a row.

☐ Establishing a uniform presentation template for each of the design teams to use when presenting their ideas will be helpful. List the number, types, and sizes of required drawings or sketches, requirements for written material, and key points for them to address in a PowerPoint presentation.

☐ At-large team members can visit each design team during their design workshops; however, they should not discuss the other teams' ideas within their own team. Usually, sponsors, project managers and specialists form the at-large team.

☐ The design teams should be supplied with area maps, Internet access and drawing supplies.

☐ Each design team should include at least one person capable of drawing and at least one person capable of creating PowerPoint slides.

☐ Limit the time for the design teams to work. Establish a strict schedule so that each team has equal advantage in developing their work. Establish the exact time that design teams must all finish, freeze, and hand in their work.

☐ Once the design teams have completed their work, set up a full-day agenda for the design review and discussion of the three design concepts. Invite interested parties to attend and discuss the presentations.

☐ Record all comments made about each of the presentations.

☐ Select the best attributes of each presentation that can be combined into the final design.

☐ Celebrate the team accomplishment!

After the three presentations have been made, the design development team should incorporate the best of the three schemes into one new scheme.

As it is still early in the project at this point in planning, the design development team should seek to create options rather than limit them. Options should be clearly expressed and numbered for reference. They should outline specific options that the Energy Conservation Plan and Renewable Energy Plan should study and develop in Steps 3 and 4.

The design development team should initiate specific research tasks and hand them off to available teammates for resolution. The design team should then make presentations of their progress during Step 2. It is now time to proceed to Step 3.

Notes

# ENERGY CONSERVATION PLAN

## *Step 3 of 10*

The third of the ten steps begins with a more developed ZEC planning team and a completed Design Development Document in place. During this step, the ZEC planning team will focus on the opportunity for energy conservation in the ZEC, addressing energy used for transportation, buildings, and other uses.

The foundation of any ZEC is the conservation of energy. It is extremely important to realize as much energy conservation as possible, because in subsequent steps, the planning team will plan enough renewable energy supply to meet all the demands of the community. It is inevitable that there will be space and cost limits to how much renewable energy equipment can be provided for the ZEC, so it is well to keep in mind that generally speaking, it is less expensive to save energy than to produce and utilize energy.

### *Craft the ZEC Energy Conservation Plan*

(1.) Review the Design Development Document with the planning team, paying particular attention to:

    A. Number and types of buildings

    B. Building sizes, ages and conditions

    C. Site exposure to wind

    D. Site exposure to solar energy

    E. Opportunities for windbreaks and/or shading

(2.) Develop ideas for building energy efficiency opportunities for electricity and thermal energy based on a rough outline of the conservation plan determined in Step 2 and consideration of:

    A. Building designs and building orientation to the path of the sun

    B. Local climate

    C. Detrimental solar exposure

    D. Wind direction and speed

    E. Seasonal air temperature data

    F. Landscape design for trees

(3.) Develop detailed information regarding energy conservation technology, operational, and financial considerations/standards:

    A. Understand the best options for energy conservation within the ZEC layout that are appropriate green building standards for:

        i. New commercial buildings

        ii. New residential buildings, by type

        iii. Attached housing

        iv. Detached housing

        v. Multi-family housing

   B. Determine the available incentives and financing programs related to energy efficiency projects:
- i. Purchase programs, rebates and tax credits associated with purchases of equipment, material and services for HVAC, water heating, insulation, tree-planting
- ii. Contact the local Governor's Energy Office (GEO).
- iii. Contact local government sustainability departments.
- iv. Contact the local utility companies (power, water, gas).
- v. Contact the DOE Energy Star Program.
- vi. Contact the DOE Energy Efficiency and Renewable Energy Program.

   C. Determine the cost premium for achieving energy efficiency for each class of building and the marketability factors for green buildings:
- i. Discuss with commercial developers.
- ii. Discuss with local/national homebuilders.
- iii. Discuss with Energy Performance Contractors.

(4.) Incorporate new team members and expertise on the planning team.

(5.) Understand any additional analysis required.

(6.) Provide further validation of the project timeline.

(7.) Update Design Development Document with new information.

(8.) Deliverables: Updated Design Development Document to include new section called Design *Guidelines for Energy Conservation.*

(9.) Meetings: Leadership, team development, planning track teams, planning workshops, research, group meetings.

(10.) Other activities: Team building, obtain professional opinions on weatherization and building costs, desk review, research, field trips.

(11.) Outcomes: Team and project design development.

(12.) Duration: 45-60 days.

(13.) See other Sections: Financial Considerations – Zeconomics (Section 3), How We Use Energy (Section 5), Conservation (Section 6), Electric Vehicles (Section 10), Other Programs and Standards (Section 14)

### Key Questions for the ZEC Team

(1.)   What information about sustainable site development is required?

(2.)   What information about green building is required?

(3.)   What examples of energy conservation can the team evaluate?

(4.)   What energy efficiency standards for newly constructed residences and buildings will be required?

(5.)   Will energy efficiency standards be mandated through deed covenants?

(6.)   Will energy efficiency renovations to existing structures be mandated?

(7.)   What will trigger the energy-efficient renovation of existing structures?

(8.)   What will be the requirement for energy-efficient appliances?

(9.)   What programs will be deployed to educate and encourage ZEC occupants to reduce energy use?

(10.)  What energy conservation plans related to transportation can be made for the ZEC?

(11.)  How will energy be conserved within the public domain?

(12.)  How can the ZEC ensure acceptable weatherization or construction costs for all ZEC occupants?

(13.)  Will a building department, homeowner's association or other entity administer the community's energy conservation?

(14.)  What people and skill sets should be added to the team?

(15.)  What other local organizations, agencies, firms and people are synergistic with planning the conservation aspect of the ZEC?

(16.)  What additional consultants are required?

Notes

*Activities for Conservation Planning*

(1.)  As a group, discuss goals for energy conservation.

(2.)  Review trends in sustainable site development.

(3.)  Review trends in green building design construction and costs.

(4.)  Make site visits to see conservation techniques and peer projects.

(5.)  Identify ideas for reducing energy use by occupants (e.g., eliminating plug-loads, using clothes lines, turning off lights, wearing slippers indoors).

(6.)  Temporarily dissolve the track teams. Appoint three team leaders for the following energy conservation sub-teams if these categories are applicable:
A.  New building energy (commercial and residential buildings)
B.  Old building energy (commercial and residential buildings)
C.  Public Domain (addresses streetlights, irrigation, transit stops)

(7.)  Each of the three sub-teams should present a recommendation for green building and other standards to be used by the ZEC, then the whole group can discuss and combine those results into a single recommendation.

(8.)  Have each of the three create a recommendation for the ZEC energy conservation goals.

(9.)  Determine the community's total requirement for energy to heat, cool, light, and power buildings based on varying building envelope standards.

(10.) Add content to update the Design Development Document.

(11.) Complete an energy conservation report for use in the next step – Creating a Renewable Energy Plan.

(12.) Energy Conservation Deliverables: The deliverables for Step 3 include three new Design Guideline documents for new and existing buildings, as well as public domain. Each Design Guideline will be incorporated in the Design Development Document describing conservation planning, including:
A.  Establish campus design standards (e.g., LEED ND)
B.  Building envelope standard (e.g., Household Energy Rating / Energy Star)
C.  Applicable energy performance testing methods and requirements
D.  Expectations for renovation of existing structures
E.  Utility systems in the public domain (e.g., solar streetlights)
F.  Mass transit accommodation in the ZEC
G. Plans for walking and bicycle paths
H. Local business development initiatives. (e.g., locate a grocery onsite to reduce traffic)
I.  Recommended programs to encourage or enhance energy conservation, (e.g., incentivize purchase of electric vehicles)
J.  List of additional requirements
K. Detailed timeline and future activities for implementing conservation
L.  Outline of preliminary plans for programs, education, and communications outreach related to energy conservation:
    i.   Training programs
    ii.  Goal setting
    iii. Information systems for tracking
    iv.  Incentives

M. Update Executive Summary to address the ZEC project's goals for conservation of energy.

The Design Development Document should be drafted and updated continuously throughout each of the ten steps. This document will describe the information known about the project requirements along each project track. The document should reflect new options and additional information requirements at the end of each step.

Analysis by the Finance team can emerge during this step. That team can forecast financial projections that compare the returns on investments for conservation in comparison to a status-quo baseline and local grid-parity energy cost.

The Conservation section will be added to the Design Development document in this step. Various maps, plans, sketches, and photos should be organized to further explain the design intent. Large format plans, blueprints and similar visuals should be presented along with informative narrative within the Design Development Document. These elements are expressions of intention, illustrations of planning studies, and tables of information. Such documents and drawings typically do not describe how buildings, systems, or landscape features are actually built. The focus of this step is defining the conservation performance attributes and creative ideas for achieving that performance and measurably good results.

When finished with this step, celebrate the teams' accomplishments and begin Step 4.

Notes

*Resources*

| ORGANIZATION | WEBSITE |
| --- | --- |
| DSIRE is the most comprehensive source of information on incentives and policies that support renewables and energy efficiency in the United States. | http://www.dsireusa.org/ |
| USGBC (2009) LEED Neighborhood Development Rating System, Washington DC: Congress for the New Urbanism, Natural Resources Defense Council & The US Green Building Council: | http://www.usgbc.org |
| United States Department of Energy—Energy Star Program and Energy Efficiency and Renewable Energy (EERE) | http://www.energy.gov |
| Passive House Institute US (PHIUS) | http://www.passivehouse.us |
| California Energy Commission—HERS– (Home Energy Rating System) | http://www.energy.ca.gov |
| United States Environmental Protection Agency (EPA), community water | http://www.epa.gov/watersense |
| Institute for Sustainable Infrastructure | http://www.sustainableinfrastructure.org/ |
| Design with Climate: Bioclimatic Approach to Architectural Regionalism: (Olgyay and Olgyay 1963) | Guide book (Available on Amazon.com) |
| National Renewable Energy Laboratory | http://www.nrel.gov |
| Governor's Energy Office | http://www.fullname of state.gov/energy/ |

*Figure 33 - RESOURCES FOR ENERGY CONSERVATION*

## *Step 4 of 10*

The fourth step of the ten steps begins with a more developed ZEC planning team and an updated Design Development Document that includes a study of the ZEC's requirements with thermal and electrical energy included. Information about building areas, policies, and the ZEC layout is also available by this stage in the process.

At this point, the ZEC regulation-track planning team should have gathered information and resources about the laws and programs addressing renewable power in their state. In this step, the possibilities are explored to determine the mix and sources of ZEC renewable energy to be utilized, including specific technologies, onsite and offsite renewable energy options, and a possible mix of those sources.

The renewable energy plan will determine how to provide renewable energy to the ZEC buildings and vehicle-charging infrastructure in the ZEC. The calculation for energy required for transportation addresses the power distribution for electric vehicles at all parking places within the ZEC. The energy for passing vehicles, deliveries, and large transit is not included in the ZEC's energy requirement.

There may be space and cost limits affecting how much renewable energy can be provided for the ZEC. The renewable energy plan may be a phased program that continually closes the gap to net-zero energy. The goal of the ZEC is to achieve a net-zero balance of supply and demand for electricity and/or thermal energy. The goal is to provide the renewable energy supply sufficient to meet all the demands of the community; these may be met by any of several renewable energy sources, providing electricity or heat, individually or in a combination. These resources may be onsite, offsite, or virtual renewable energy credits, or a combination of two or more.

### *Objective: Craft the ZEC Renewable Energy Plan*

(1.) Incorporate new team members and expertise on the planning team.

(2.) Share the information in the Design Development Document and the Conservation Plan.

(3.) Understand the regulations affecting the generation, distribution and sale of electricity and thermal energy in your state and locale.

(4.) Determine whether the ZEC will operate on-grid or off-grid and what the:

  A. Average annual requirement for renewable energy needs to be to achieve net-zero operation.

  B. Peak energy requirement will be if the ZEC is to be off-grid, or incorporate emergency power backup.

  C. Dispatchable energy generation requirement will be in order to supplement the non-dispatchable energy (e.g., solar/wind) – how to support peak energy required by an off-grid ZEC.

  D. Examine the ZEC's time-of-supply and time-of-use scenarios.

  E. Determine if there will be any requirement for backup generators, considering:

   i. Size, number and location of generators

      ii. Requirements for connection to any fuel pipeline

      iii. Area for storage of fuels onsite

      iv. Noise and vibration

      v. Environmental and air permitting requirements

(5.)  Perform a screening to identify and quantify all possible sources of renewable power and thermal energy available for the ZEC:

    A. Onsite energy

    B. Offsite energy (e.g., solar garden)

    C. Renewable energy supplies available over the power grid

    D. Renewable energy credits (RECs)

(6.)  Develop detailed information regarding energy conservation and technological, operational, and financial information regarding site-appropriate renewable energy technologies:

    A. Determine energy cost and projected cost escalation.

    B. Determine financial incentives such as rebates, tax credits, PACE finance, etc.

    C. Determine renewable energy financing sources and evaluate programs.

    D. Examine options for developing energy solutions with outsourced options from Energy Service Companies (ESCO) and/or utility companies (buy/build analysis).

(7.)  Understand all of the best options for renewable energy generation within the ZEC layout.

(8.)  Understand the ZEC's key possibilities for renewable energy features and benefits.

(9.)  Understand cost and financing implications for renewable energy.

(10.) Understand any additional analysis required.

(11.) Provide further validation of the project timeline.

(12.) Update Design Development Document with new information.

(13.) Deliverables: Updated Design Development Document to include new section called Design Guidelines for Renewable Energy.

(14.) Meetings: Leadership, team development, planning-track teams, planning workshops, research, group meetings.

(15.) Other activities: Team building, obtain professional opinions on renewable energy and related building cost, desk review, research, field trips.

(16.) Outcomes: Team development, Renewable Energy Plan, input regarding policies.

(17.) Duration: 45-60 days.

(18.) See other Sections: Financial Considerations – Zeconomics (Section 3), How We Use Energy (Section 5), Exploring Renewable Energy (Section 9), Electric Vehicles (Section 10), Other Programs and Standards (Section 14).

*Key questions for the ZEC Team*

(1.)  Is there adequate information available about the ZEC's overall energy requirements for both electrical and thermal energy?

(2.)  What renewable energy is available for purchase on the grid, or is legally possible when wheeled through the grid (accomplished via net metering or virtual net metering), or available through programs for solar garden, renewable energy credit, and/or other programs?

(3.)  What financial incentives affect the economics of the renewable energy plan?

(4.)  What are the siting options for onsite and offsite renewable energy from wind, solar, hydroelectric, geothermal, and tide?

(5.)  What portion of the buildings' roof areas have potential for solar access, and how many watts could be produced from that?

(6.)  What other site areas are suitable for solar access, and how many watts could be produced from them (e.g., solar garden, solar parking canopies)?

(7.)  What other information about renewable energy is required?

(8.)  What examples of renewable energy can the team evaluate?

(9.)  Will renewable energy equipment additions to existing structures be mandated?

(10.)  What will trigger renewable energy additions to existing structures?

(11.)  What renewable energy equipment will be required for newly constructed residences and buildings? Will they be mandated?

(12.)  What programs will be deployed to educate and encourage ZEC occupants to use renewable energy?

(13.)  How can the ZEC ensure acceptable renewable energy cost for all ZEC occupants?

(14.)  Will a building department, homeowner's association or other entity administer the community's renewable energy program?

(15.)  What people and skill sets should be added to the team?

(16.)  What other local organizations, agencies, firms and people are synergistic with planning the conservation aspect of the ZEC?

(17.)  What additional consultants are required?

*Activities for Renewable Energy Planning*

(1.)  As a group, discuss goals for renewable energy.

(2.)  Review trends for renewable energy utilization in sustainable development.

(3.)  Review trends in supplying and managing energy in green buildings.

(4.)  Develop an understanding of methods for cost-modeling of energy.

(5.)  Make site visits to see renewable energy installations and peer projects.

(6.)  Develop a renewable energy plan that indicates the equipment locations.

(7.)  Develop a table of all sources and uses of energy.

(8.)  Develop a design guideline for renewable energy to be used by the ZEC.

(9.)  Add content to update the Design Development Document.


*Renewable Energy Plan Deliverables*

(1.)  The deliverables for Step 4 include updates to the Design Development Document and should include the following:

    A.  A summary of plans for each type of renewable energy to be used, where the equipment is located, the annual production expected, and how it will be connected, or virtually connected.

    B.  The plan for phasing-in renewable energy supply capacity over the life of the ZEC.

    C.  The plan for offsite renewable energy from new or existing sources. (include maps and preliminary equipment siting plans)

    D.  The plan and drawings for the onsite renewable energy equipment, including existing and planned installations.

    E.  A draft of the Design Guideline for renewable energy, addressing all requirements for new and existing buildings, and public domain. This should explain expectations for renewable energy equipment installations for new and existing structures (e.g., 2.5 kW solar minimum).

    F.  A plan for utility systems in the public domain (e.g., solar streetlights).

    G.  A plan for energy generation performance and testing describing responsibilities, methods and reporting requirements

    H.  A plan for local business development initiatives (e.g., locate a brewery onsite and use waste heat).

    I.  Detailed timeline and future activities for implementing renewable energy.

    J.  Outline of plans for education and outreach related to renewable energy uses:

        i.  Training programs

        ii.  Goal setting

        iii. Information systems for tracking

        iv. Incentives

    K.  List of additional requirements

    L.  Executive Summary update addressing the ZEC project's goals for renewable energy

The Design Development Document should be drafted and updated continuously throughout the process of each of the ten steps. This document will describe the information known about the project requirements along each project track. The document should reflect new options and additional information requirements at the end of each step.

Analysis by the Finance team can emerge further during this step. That team can demonstrate sensitized financial projections that model the returns on investments for renewable energy utilization in comparison to local grid-parity energy cost.

Various maps, plans, sketches and photos should be organized to further explain the design intent. The large format plans, blueprints, maps, and diagrams should be presented along with information written in the Design Development Document.

The design development documents and drawings are usually expressions of intention, illustrations of planning studies, and tables of information. Design development documents and drawings usually do not describe how buildings and systems or landscape features are actually built. The focus of this work is defining the renewable energy installations, connections and related systems such as backup generation. Include performance attributes and creative ideas for achieving that performance, and measurably good results.

When completed with this step, celebrate the teams' accomplishment and move to the next step.

Notes

*Resources*

| ORGANIZATION | WEBSITE |
| --- | --- |
| DSIRE is the most comprehensive source of information on incentives and policies that support renewables and energy efficiency in the United States | http://www.dsireusa.org/ |
| Department of Energy (DOE) Qualified List of Energy Service Companies (ESCOs) (DOE Qualified List) in accordance with the Energy Policy Act of 1992 (EPAct1992) and 10 CFR 436 | http://www1.eere.energy.gov/femp/financing/es-pcs_qualifiedescos.html |
| United States Department of Energy—Public Training and Sunshot Community Solar Development Program | http://energy.gov/publicservices/energyeconomy/education-training |
| Passive House Institute US (PHIUS) | http://www.passivehouse.us |
| California Energy Commission—HERS– (Home Energy Rating System) | http://www.energy.ca.gov |
| United States Environmental Protection Agency (EPA), air permitting of generation equipment | http://www.epa.gov/watersense |
| National Renewable Energy Laboratory: Access resources for determining available solar, wind, geothermal, and hydroelectric energy | http://www.nrel.gov |
| Governor's Energy Office: Access resources on regionalsolar, wind, geothermal and hydroelectric energy. | http://www.fullname of state.gov/energy/ |

*Figure 34 - RESOURCES FOR RENEWABLE ENERGY PLAN*

## *Step 5 of 10*

The fifth step of the ten steps begins with a more developed ZEC planning team and an updated Design Development Document that includes a study of the ZEC's requirements and potential supplies for thermal and electrical renewable energy.

During this step, the ZEC planning team will explore the financial, implementation and operational aspects of the ZEC. Regardless of the confidence level of the ZEC team, achieving their community goal requires that the solutions be technically possible and economically and operationally feasible. The voice of the marketplace needs to be understood. The competitive dialogue process provides the opportunity for the ZEC team to explore the capabilities of equipment and service providers to meet the requirements of the ZEC.

Whether the ZEC needs a solar integrator with financing or an energy service company to provide a complete micro-grid, the competitive dialogue procurement process is valuable. This is an emergent process, informed not only by the ZEC teams' initial understanding of their business and technical requirements, but also information about the unique and often valuable methods that vendors may use to provide solutions that meet those requirements.

Procuring the equipment and services needed for the ZEC involves participating in a fast-growing marketplace associated with energy in the US. This energy business sector has been burgeoning in the US, and is therefore maturing. The technologies involved continue to improve in performance, and costs are being reduced while financing alternatives expand. Because of the changes occurring in the energy industry, ZECs will face challenges in understanding what products and services are too old or too new for them.

There may be a need for technically complex or logistically difficult solutions to be implemented, especially where the construction or renovation of buildings or energy systems is required. In order to achieve the completion of the ZEC and reasonably protect the community from exposure to financial, technological, or execution risks, a competitive dialogue process is recommended. This process involves an early engagement of qualified suppliers competing to provide the best value for the ZEC.

Executing a competitive dialogue process includes the identification of ZEC requirements, vendor sourcing and contracting. The qualification of suppliers requires research, briefings, and the development of selection criteria. ZEC planners need to break the packages out according to a logical program that is consistent with what the suppliers in the market offer. Developing and packaging the procurement documents, drawings, and response forms usually leads to answering vendor questions and vendor presentations. Solicitation of pricing proposals, technical and commercial evaluations follow, and then negotiation and contracting are next.

The competitive dialogue process is iterative, informative, and precise in guiding procurement decisions for the benefit of the ZEC. Enhancement of the proposed solutions and price advantages may occur at each step in the process. The process progresses through the steps described in this following section, providing an opportunity for value-realization through a structured communication between the ZEC and the bidders.

### ZEC Requirements for the Competitive Dialogue Process

(1.) Competitive Dialogue Deliverables: The deliverables for Step 5 include updates to the Design Development Document, which describe:

    A. The broad scope of the procurements required for the business partnerships, services and equipment required to fulfill the ZEC program. This may be to procure an agreement with any combination of the specific procurement needs for the ZEC for the following and other requirements:

        i.  Homebuilder(s) to design and construct new homes

        ii.  Developers of office, retail, condominium or apartments

        iii.  Wind turbines, solar panels, electric vehicle charging stations, generators, control systems, or other systems

        iv. Geothermal energy well-drilling services, or loop-piping system construction for district heating

    B. The scope of the Competitive Dialogue Process

    C. Identification of key supply requirements for the ZEC, determining the individual procurement packages required if more than one

    D. Other procurement matters required by the ZEC business entity and for purchase programs on behalf of the ZEC occupants

    E. Expectations for the management of the procurement process

(2.) Identify potential suppliers to the ZEC

(3.) Advertise the requirement and provide a Request for Information (RFI) that invites the formation of vendor consortiums, and delineates timing, and specific information to be included in vendor's Statements of Qualifications (SOQs)

(4.) Determine which vendors are compliant with the commercial and technical requirements and which are qualified bidders for this ZEC project.

(5.) Bind the vendors' organizations to agreements that require appropriate agreements to non-disclosure and prevent the ZEC from obligation to make contract awards, reimburse vendor costs, and protect the ZEC from any other risk.

(6.) Compile an Executive Brief presented to all vendor groups simultaneously describing the vision, mission, goals and general procurement requirements for the ZEC.

(7.) Arrange individual vendor presentations to introduce the prospective teams, capabilities, and recommended solutions for the ZEC. Provide the opportunity for the ZEC team to interview vendors.

(8.) Receive and answer vendors' requests for information (RFIs) regarding requirements, preferences, and procedures.

(9.) Analyze needs for the ZEC in context of supplier capabilities and group procurements into separate packages as required. (e.g., separate bids for professional services, energy services or equipment, green building construction, weatherization).

(10.) Develop preliminary, draft and final versions of procurement requests for the services and equipment required in fulfilling the ZEC program on behalf of the community, or as a program for all individual members of the community to partake, or both.

(11.) Receive preliminary proposals and evaluate proposal compliance with requirements. It is recommended that technical and commercial information be kept separate and that an information firewall be made between technical and financial evaluators.

(12.) Arrange vendor sponsored visits to peer projects, factory tours, product demonstrations, virtual tours and other informative research for the ZEC team, and document the findings.

(13.) Provide specific feedback to each bidder about areas where their proposals are not compliant with requirements, and then request a best and final offer.

(14.) Evaluate best and final offers, and enter into negotiation simultaneously with all competing bidders. Continue to provide the opportunity for them to improve their offers through a dialogue focused on value-realization for the ZEC.

(15.) Appoint vendors and negotiate the details of the scope of work, performance metrics and standards, schedules, project communications, warrantees, service-level agreements, bills of materials, terms and conditions, change-orders, use and venue of law. (Determine the implications a force majeure, which includes performance during or after a natural disaster, war, strike or other unavoidable interference.)

(16.) It is recommended that the ZEC team organize the project kick-off activities, and appropriately celebrate the appointment of key business partners and vendors.


### Objectives of the Competitive Dialogue Process

(1.) Secure dependable solution providers to meet the precise needs for all requirements of the ZEC project, at the lowest cost.

(2.) Assure technical compliance for all equipment and service solutions, based on specific performance criteria and the ZEC's intended results.

(3.) Identify qualified suppliers for the equipment and services required for the ZEC.

(4.) Bundle vendor responsibilities to deliver complete integrated solutions that deliver required performance levels.

(5.) Identify the details for vendors' scope of work, including design, engineering, fabrication, delivery, installation, testing, training, and documentation.

(6.) Provide the ZEC team with an understanding of all related costs, value-engineering alternatives, optional functionality, financing options, phasing options, and the range of services available and their cost.

(7.) Understand the schedule and work sequence for implementing conservation and renewable energy projects at the ZEC.

(8.) Determine the most appropriate Service Level Agreements (SLAs) and key performance indicators (KPIs) to insure functionality, service call and repair responses, and warrantee and customer service obligations of vendors.

(9.)  Assure holistic project responsibilities through integrated delivery of building services, equipment supply, and other services provided by vendors that are made responsible for overall functionality, assurance of system compatibilities, and overall system integration and operation.

(10.)  Reasonably assure that solutions required, and proposed solutions, are actually constructible, buildable, and practical.

(11.)  Promote ethical business behavior and prevent corrupt business practices.

(12.)  Deliverables: Various documents and document reviews used in the solicitation and evaluation of proposals and appointment of vendors and contracts known as RFIs, SOQs, RFPs, RFIs, NDAs, technical compliance and commercial evaluations, reports, letters of intent or understanding, terms sheets, bills-of-materials, service-level agreements, power purchase agreements, and purchase orders.

(13.)  Meetings: Leadership, team development, finance team, planning-track teams, briefings, vendor presentations and workshops, research, group meetings, site visits, mockups, etc.

(14.)  Other activities: Financial and technical analysis, legal review

(15.)  Outcomes: Team development

(16.)  Duration: 60-180 days

(17.)  See other sections: Financial Considerations – Zeconomics (Section 3), Conservation (Section 6), Exploring Renewable Energy (Section 9)

## Notes

*Key Questions for the ZEC Team*

☐  How will the ZEC team manage the integrity of procurement processes in a manner that protects the ZEC's interests?

☐  How can the ZEC team avoid procuring equipment systems or services that do not perform as required?

☐  How do the ZEC vendors appropriately share financial risk and rewards with the ZEC?

☐  What are the general qualifications required for each vendor to be considered by the ZEC?

☐  What are the appropriate packages to be bid, and can a large vendor or consortium bid multiple packages?

☐  How will the technical team make their evaluations—on technical merits alone?

☐  How much work can the vendors be expected to do before they are put under contract?

☐  How is team chemistry evaluated in making purchase agreements?

☐  What items being purchased are most unique, new, or unusual to the ZEC planning team and how can those elements be demonstrated?

☐  How can the ZEC leverage the procurement process to learn the most about the technical, operational, legal/regulation, marketing and project management aspects of their ZEC?

☐  What will be the process for sourcing vendors?

☐  Will local vendors be given a preference in making contract awards? How much (e.g., 5% to 10%)?

☐  How can the team dedicated to the finance track present the summarized cost evaluations in a manner that assumes incentives, addresses life-cycle costs, and otherwise levels the bids against the actual requirements?

☐  Does the ZEC team appreciate the need to take procurements through stages by asking for initial estimates, bids, best and final offers (BAFO's)?

☐  How will the lessons learned during the competitive dialogue process be integrated with the six planning tracks?

*Activities for the Competitive Dialogue Process*

(1.) Expand the net of the ZEC planners by reaching to the broader community to support the competitive dialogue process.

(2.) Establish connections with local chambers and other business groups that can assist in the ZEC's competitive sourcing.

(3.) Discuss the needs for the ZEC.

(4.) List the ZEC's needs for:
   A. Energy conservation
   B. Renewable energy
   C. Accommodation of EV charging
   D. Renewable energy supply

(5.) Discuss the list above and define the unique needs in each category in accordance with the Design Development Plan.

(6.) Determine what information or expertise is required to provide the team with confidence in each area of procurement.

(7.) Review trends in supplying and managing energy to communities.

(8.) Research procurement trends for ZECs and eco-cities developments.

(9.) Develop an understanding of energy cost-modeling methods.

(10.) Make site visits to see installations and peer projects; ask about vendors.

(11.) Determine the highest levels of measure that apply to each procurement, and expand that information into a description of required performance.

(12.) Develop concepts and plans for testing against the specified performance requirements and build a table for performance metrics.

(13.) Develop a preliminary procurement plan for goods and services.

(14.) Develop contacts and relationship with supply chain sources for needed packages.

(15.) Receive vendor responses containing company, product, service and other ZEC information requirements.

(16.) Develop and publish a schedule for active package procurements.

(17.) Develop a single point-of-contact structure for vendors and the ZEC.

(18.) Provide an executive briefing to orient vendors on requirements for participation in the procurement.

(19.) Qualified vendors demonstrate they are capable, technical competent, and meet business and financial requirements.

(20.) Discuss ZEC goals, achievements, and timeframe.

(21.) Review the general requirement of the ZEC and procurement package.

(22.) Discuss the applicable performance requirements for:
   A. Planning and engineering
   B. Manufacture
   C. Installation
   D. Training
   E. Service and support
   F. Engage vendors to bring a proposal:

        i.  For discussion
        ii. Then another for budgeting
        iii. Then another for their best and final offer
    G. Have the Technical Team provide a gap analysis comparing technical solutions proposed for the ZEC with the specific performance requirements.
    H. Negotiate the exact requirements, approach, and price, then:
        i.  Negotiate terms and conditions
        ii. Appoint vendors
        iii. Manage vendors

(23.) Competitive Dialogue Deliverables: The deliverables for Step 5 include updates to the Design Development Document, describing:
    A. Expectations for the management of the procurement process including:
    B. Stating the governance procedures
    C. Authorizing the ZEC financial team
    D. Designating reporting and controls
    E. Requiring budget approvals
    F. Other requirements of leadership
    G. A list of all major procurement items, broken down into groups according to supply chain sources
    H. Summary description for each of the ZEC procurement packages
    I. A schedule showing the time frame and steps for each package being procured—in alignment with the overall ZEC schedule
    J. A process document indicating how the RFQs, SOQs, RFPs, RFIs, NDAs, SOWs, SLAs, KPIs, and other supply chain processes will be administrated
    K. Testing plan addressing the methods to be used to inspect and certify the vendor's progress and deliveries at office, factory or ZEC
    L. Recommended programs to encourage achievement of supply chain goals
    M. Outline of plans for education and outreach related to procurement

(24.) The Design Development Document should be drafted and updated continuously throughout the process of each of the ten steps. This document will now describe the information known about the project requirements, budget, schedule, and performance along each project track. The document should reflect new options, decisions made, and additional information requirements at the end of each step.

Analysis by the Finance Team can solidify during this step as more information about the sensitized financial projections will be available and the economic models that demonstrate the returns on investments for the ZEC will be more accurate.

Various maps, plans, sketches and photos developed to inform or respond to the procurement competitive dialogue should be organized to further explain the design intent. The large format plans, blueprints and similar materials should be presented along with information written in the Design Development Document.

The design development documents and drawings are usually expressions of intention, illustrations of planning studies, and tables of information. The vendor's plans, submittals and drawings will describe how buildings, systems or landscape features are actually to be built. When completed with this step, celebrate and move on to Step 6.

Notes

### Resources

The private finance initiative: How to conduct a competitive dialogue procedure. LEEDs, UK: United Kingdom, Department of Health Private Finance Unit. (Katherine 2006)

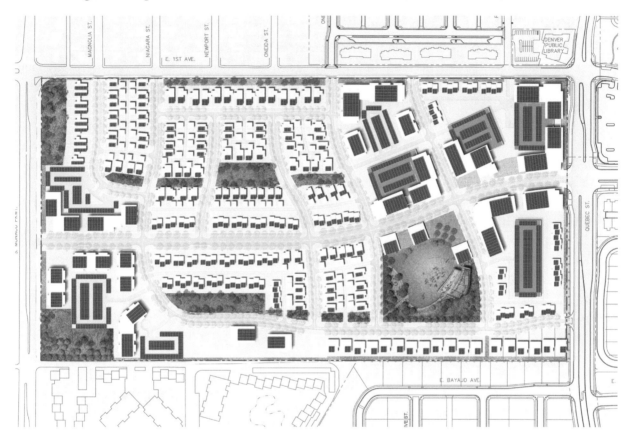

*Figure 35 - A SITE SCREENING UNVEILS THE POTENTIAL FOR RENEWABLE ENERGY. BLUE INDICATES THE POTENTIAL LOCATIONS FOR PHOTOVOLTAIC SOLAR PANELS. (Courtesy of Lowry and Design Workshop, Inc.)*

# SECTION 23
# PUBLIC DOMAIN | PUBLIC PROCESS

## *Step 6 of 10*

The sixth of the ten steps begins with a substantively developed ZEC plan in place and at least partial completion of the procurement process. In order to fix the plans for the ZEC, a number of authorities may have to make approvals, and the public will weigh in with questions and comments regarding various aspects of the development.

The requirements in this step may vary tremendously depending on the extent of new development and construction required for the ZEC. A Greenfield ZEC that includes new streets, sewers, utilities and transit stations requires significantly more attention in this step than a ZEC that is a renovation of an existing neighborhood.

During this step, the ZEC planning team will address the needs of all authorities having interest or jurisdiction at the site, develop acceptable plans, and gain approvals for annexing, tax and energy district formations, traffic planning and parking, and entitlement for the development. Additionally, the ZEC will require coordination with departments such as public works, water and wastewater, transit authorities, building departments, fire marshals, and local utilities.

The planning of the public domain addresses matters of environmental impact: storm water control, street construction, alleys, utility easements, rights-of-way, firefighting, and public transportation, which affects public land and operation of public services. This phase of the ZEC plan may also include requirements for building design.

All new developments require approvals from municipalities, counties or state authorities. ZEC planners may find local landscape design firms, architects, and engineers helpful in this phase, as they may be familiar with the agencies, people and processes involved in the entitlement of new real estate developments and permitting processes.

The public process provides opportunity for citizens and other institutions and businesses to examine the General Development Plan (GDP), to ask questions and to register their comments for consideration by the authorities evaluating applications for zoning or development.

> *Recognizing that entitlement authorities represent customers to be sold on a project, experienced developers have learned that it is useful to address local authorities' concerns from the onset. A series of negotiations often transpires as developers seek to tailor their projects to regulators' expectations. It is far better to identify and address community concerns early in the project approval process than to face an anxious audience in a public hearing. Elected officials are much more comfortable issuing approvals when the electorate is at ease with a project. (Brett and Schmitz 2009) – Urban Land Institute*

### *Objectives of the ZEC Public Domain and Public Process*

(1.)  Develop an assessment of the economic impact of the ZEC.

(2.)  Provide municipal annexation, if required or beneficial to the ZEC.

(3.)  Achieve all entitlement and permitting requirements for all ZEC land uses.

(4.)  Secure tax incremental financing (TIF) commitments for redevelopment or infrastructure and other community improvements.

(5.)  Achieve desired zoning designations for the ZEC property.

(6.)  Establish tax districts, if required.

(7.)  Establish energy districts, if required.

(8.)  Present the ZEC general development plan (GDP).

(9.)  Facilitate public review, public comment, and resolve public concerns.

(10.)  Conclude environmental assessments that may be required for the ZEC.

(11.)  Obtain compliance with recommendations of traffic studies.

(12.)  Approve layout of streets, walking trails, lot lines, sidewalks, easements, alleys and utility layouts.

(13.)  Approve plans for parks and recreation.

(14.)  Establish all site requirements for firefighting, including clearances, red-curb fire zones and water hydrants.

(15.)  Seek approval of any ZEC ordinances required to ensure solar access.

(16.)  Seek approval for use of unusual equipment such as geothermal district loop fields, electric vehicle charging stations used in public parking spaces, and/or solar streetlights.

(17.)  Approve plans for historic or cultural heritage preservation at the site.

(18.)  Coordinate detailed requirements for ZEC design with all agencies and utilities with authority or interests related to the ZEC site:

    A. Address storm water management, riparian systems, required pervious surfaces and storm water detention basins and culverts.

    B. Determine how mass transit authorities will serve the ZEC development. Coordinate all requirements for road crossing easements, passenger stations, and parking facilities.

    C. Requirements of the water and/or sewer providers for the ZEC, addressing potable and recycled sanitary irrigation water plan.

    D. Early discussions with building departments and the fire marshall will foster collaboration and cooperation regarding the unique aspects of the ZEC. (e.g., fire authorities may require controls for electrical power turn off (PTO), roof access and venting areas on buildings)

    E. Determine requirements for installation or modification of electric and gas utilities or telecommunications utilities.

    F. Coordinate requirements for snow removal, street sweeping, waste management and postal delivery.

    G. Determine a plan for crosswalks, signals and school zones.

(19.) Incorporate new team members and expertise on the planning team.

(20.) Understand any additional analysis required.

(21.) Deliverables: General development plan (GDP), applications, licenses, filing fees, and updates to Design Development Document.

(22.) Meetings: Hearings, public meetings, coordinating meetings with agencies, leadership, team development, planning-track teams, planning workshops.

(23.) Other activities: Team building, obtain professional opinions, desk review, research, field trips.

(24.) Outcomes: Entitlement of ZEC

(25.) Duration: 45-60 days

(26.) See other sections: Financial Considerations – Zeconomics (Section 3), Twenty-five Objections to a ZEC (Section 4), How We Use Energy (Section 5), Conservation (Section 6), Electric Vehicles (Section 10), Other Programs and Standards (Section 14).

### *Key Questions for the ZEC Team*

(1.)  What expertise does the ZEC planning team require to address the public domain and public process?

(2.)  What specific site development approvals are required?

(3.)  What permits are required for individual lots and buildings in the ZEC?

(4.)  Will the ZEC be seeking financial assistance, such as tax-increment financing?

(5.)  Who are the agencies and entities involved in the entitlement, permitting, and planning process?

(6.)  Has the planning team addressed all of the concerns in the design?

(7.)  Who will prepare the general development plan (GDP)?

(8.)  Who will prepare the economic development analysis, if required?

(9.)  Who will prepare any necessary environmental assessment?

(10.)  Who will address the public on behalf of the ZEC?

(11.)  Will a public relations campaign be required?

(12.)  What will necessary permits cost?

(13.)  What are likely to be the objections to the ZEC?

(14.)  How long will the process take?

*Activities for the ZEC Team*

(1.) As a group, discuss goals and challenges related to entitlement and permitting.

(2.) Develop statements describing the ZEC value proposition to the community at large.

(3.) Determine how the ZEC leadership can serve as spokespersons to smooth the entitlement process with officials.

(4.) Identify successful peer projects for use in comparison.

(5.) Explore all aspects of economic development potential and effects to the community tax revenues (e.g., increased property values, job creation, material and equipment sales, EV sales, etc.)

When completed with this step, celebrate your team's accomplishments and begin Step 7.

Notes

# SECTION 24

# ESTABLISH ZEC GOVERNANCE

## *Step 7 of 10*

The seventh step begins with a well-developed plan in place for the Zero Energy Community and a critical need to assess and determine the governance of the ZEC entity. The requirements of this step may vary tremendously depending on the extent of new development, the number and types of users who will own property in the ZEC, and the relationship of the ZEC to other development and governing authorities, agencies, and local regulators.

It is conceivable that the area in which the ZEC is being developed has existing governance mechanisms to respond to the needs of the ZEC planners, or that those entities would cooperate by adjusting their governance in order to meet the needs for the ZEC.

As discussed in Section 23: Public Domain | Public Process, the planned ZEC may include new or modified streets, sewers, utilities and transit stations. The ZEC plan will require coordination with government authorities and planning departments, such as public works, water and wastewater, transit and local utilities. The bodies who decide on zoning and district formation may have important decisions to make about the land uses and acceptable development of the ZEC.

Once commenced, the ZEC needs to be led and represented, and may need to establish authorities to act on behalf of its constituents. This step is focused on the development of the required ZEC governance to meet the specific needs of the ZEC.

The governance of a ZEC is comparable to a homeowner's association (HOA). There is a need for community representation, enforcement of rules, financial transactions amongst occupants, governance and external parties. The similarities between the ZEC and the HOA converge with a community interest to protect—including property owner and community representation, as well as common property and services to manage.

Conversely, the leadership, innovation, financial and technical expertise required for a ZEC far exceed the capacity and operating capabilities required for a HOA. In one model, it may be possible to develop an organization to handle some of the governance of the ZEC and to leave other responsibilities to a homeowner's association or to reconceive the HOA. These matters all require analysis, effective communication, and formation of a strategy.

Until such time when practices for governing ZECs are broadly developed, it is recommended that ZEC planning teams compare their requirements with those common to HOAs and add specific changes to address specific problems.

During this step, the ZEC planning team will continue to uncover and address the needs of all authorities having interest or jurisdiction at the site, developing acceptable plans, and gaining approvals for annexing, tax and energy district formations, traffic planning and parking, and entitlement for the development, along with establishing other requirements regarding energy use and generation.

### Crafting a ZEC Governance Plan

(1.) Coordinate governance with other agencies that have interests or jurisdiction.

(2.) Resolve required changes (or exceptions) to laws and regulations that are necessary in order to govern the special concerns of the ZEC, or that conflict with the goals for the ZEC (e.g., a law that bans outdoor clothes-drying may conflict with an energy-related goal of the ZEC).

(3.) Indicate the program development priorities required for the ZEC and address them through assignment of authority and responsibility to existing groups or a new ZEC governance entity.

(4.) Address the development of financial resources for the ZEC. This may inform the best form of any new entity.

(5.) Establish the governance's breadth of responsibility and authority regarding relationships between the community and other regulators and agencies.

(6.) Establish the governance's breadth of responsibility and authority regarding contractual relationships between the community and third parties.

(7.) Determine the governance's authority and plans regarding making assessments and charging fees to ZEC occupants.

(8.) Determine whether the governance must pursue formation of:

A. A tax district

B. A political district

C. An energy district

D. Municipalization

E. Annexation

F. New town incorporation

(9.) Determine whether the ZEC will pursue any of the following:

A. Government grant(s)

B. Government loan(s)

C. Tax incentive financing (TIF)

(10.) Define the realms of control for the ZEC governance, addressing the public domain, common areas, private properties, community energy conservation, community centers, and community renewable energy as these apply.

(11.) If a new ZEC governance entity is required, decide on a structure, such as a non-profit corporation, association, authority, or a public, quasi-public or private company. Prepare a statement of values for the governance, addressing integrity, leadership, and transparency.

(12.) Forge relationships, attract expertise, enlist people and required skills.

(13.) Determine a name for each new ZEC governance entity.

*Key Questions about ZEC Governance*

(1.) Is the ZEC for a single user, such as a health science center, government center, university or corporate campus, or for a mixed-use community?

(2.) If not for a single user, what are the classifications of ZEC user groups (e.g., single v. multiple types of land uses/users, retail, commercial, residential, institutional)?

(3.) Is there a requirement to coordinate governance with other bodies thst have overlapping interests or jurisdiction, such as:

   A. Federal agencies – parks, land management, military base

   B. City or village government

   C. Economic development agencies

   D. Regional development authorities

   E. Homeowner's or neighborhood associations

   F. Building, fire, and law enforcement departments

   G. Transit authorities

(4.) What unique considerations of law and regulation affect permissible activities and energy-related equipment installations within the ZEC, such as:

   A. The right to buy or sell electricity

   B. Solar access – control of vegetation and structures obscuring solar

   C. Vehicle charging – at assigned and un-assigned parking places

   D. Clotheslines – are there energy-conserving techniques that require exemption from current regulations?

   E. Regulations addressing net metering, geothermal wells

   F. Architectural design guidelines and other rules

   G. Expediting programs for permitting energy improvements

   H. Communications programs with ZEC occupants

   I. Rules of enforcement

   J. Voter registration districts and elections

(5.) Is there a need for the ZEC to create, or enjoin, any agency to perform economic development activities aimed at attracting businesses, schools and other institutions that enhance the ZEC's sustainability (e.g., providing a tax deferral)?

(6.) What legal and financial expertise is required to develop and operate governance for the ZEC and address matters such as:

   A. Analysis and advisory services

   B. Financial calculator tools / apps

    C. Property Assessed Clean Energy (PACE) finance programs

    D. Purchasing programs established to provide rebates, tax credits or discounts

    E. Lending programs for ZEC occupants to borrow money to improve their homes' or businesses' energy efficiency—or finance the purchase of renewable energy systems.

(7.) What authorities may be required to represent community interests regarding contractual relationships between the community and third-parties in order to:

    A. Develop power-purchase agreements (PPAs) with renewable energy providers

    B. Select and appoint "approved vendors" for energy-related work upon the commonly-held or private property within the ZEC, including:

        i. System integrator and/or energy performance and weatherization contractors

        ii. Contracting for engineering, equipment, or installation or operating services

(8.) What will be required to represent the community programs for:

    A. Obtaining zoning variances, development entitlements or building permits

    B. Organizing community-based development of renewable energy

(9.) How will the ZEC facilitate assessments and other fees from the occupants?

(10.) Will ZEC governance be required to create any municipalization, annexing, or creation of a new town?

(11.) Will ZEC governance be required to form a special tax district, political district, or form an energy district, or arrange tax incremental financing (TIF) funding to pay for infrastructure or other improvements?

(12.) What are the realms of control required for:

    A. Public domain

    B. Common areas

    C. Private properties

    D. Community energy conservation

    E. Community renewable energy

(13.) Given the responses to the above, what is the best structure for the ZEC governance:

    A. Non-profit corporation

    B. Association

    C. Authority

    D. Public, quasi-public or private

(14.) Should the governance be a function of an existing entity or a new entity?

(15.) What relationships, people, and skills should be added to the team?

### Activities for Governance Planning

(1.) Determine an appropriate representation of the external community on the ZEC planning team and include the representative(s) throughout this step.

(2.) As a group, discuss general goals for the ZEC governance.

(3.) Discuss goals and values desired from the governance (e.g., leadership, integrity, courage, ethics, transparency, etc.).

(4.) If a new entity is required, evaluate possible forms, such as: governmental department, association, non-profit, corporation, and development authority. Decide whether it will be private, public or quasi-public entity.

(5.) Discuss options for filling positions for any required directors, officers or leadership team for any new or re-formed entity.

(6.) Determine any other concerns regarding the governance of the ZEC.

(7.) Review trends in ZEC governance, and their relationship with homeowners associations and how these trends may inform solutions for your group's ZEC.

(8.) Discuss the goals and ideas with community leaders and incorporate their inputs in the planning process.

(9.) Outline the three best solutions and charge individual subgroups with developing and defending each approach.

(10.) Present three plans and then merge the three into a preliminary draft.

(11.) Discuss possible names for the new ZEC governance entity.

### Deliverables for Governance Planning

ZEC Governance Deliverables: The deliverables for the governance evaluation and planning process can vary considerably. In some cases a new governance entity must be established. In other cases a ZEC will expand the role and purview of an existing entity (for instance, a homeowner's association). Depending on the particular situation, the following deliverables will be required:

    A. The official name of the ZEC organization.

    B. An organizational charter that expresses the vision, mission, goals, duties, authorities and constituents of the governance entity.

    C. A charter of values expressing the immutable attributes of leadership, integrity, courage, ethics, transparency required by the ZEC founders and community.

    D. Appoint any required directors, officers or leadership team for the new or re-formed entity.

    E. Confirmation of the community agreement to the entity formation by vote or decisions by the existing electorate.

    F. Confirmation of filing of any articles of incorporation, bylaws, and/or registrations or applications for business license(s) that may be required.

    G. Issuance of state and federal tax identification numbers and employer identification number (EIN).

Now it is time to proceed to Step 8 of 10.

*Resources*

| ORGANIZATION | WEBSITE |
| --- | --- |
| DSIRE is the most comprehensive source of information on incentives and policies that support renewables and energy efficiency in the United States. | http://www.dsireusa.org/ |
| USGBC (2009) LEED Neighborhood Development Rating System, Washington DC: Congress for the New Urbanism, Natural Resources Defense Council & The US Green Building Council: | http://www.usgbc.org |
| United States Department of Energy—Energy Star Program and Energy Efficiency and Renewable Energy (EERE) | http://www.energy.gov |
| Passive House Institute US (PHIUS) | http://www.passivehouse.us |
| California Energy Commission—HERS– (Home Energy Rating System) | http://www.energy.ca.gov |
| United States Environmental Protection Agency (EPA), community water | http://www.epa.gov/watersense |
| Institute for Sustainable Infrastructure | http://www.sustainableinfrastructure.org/ |
| Design with Climate: Bioclimatic Approach to Architectural Regionalism: (Olgyay and Olgyay 1963) | Guide book (Available on Amazon.com) |
| National Renewable Energy Laboratory | http://www.nrel.gov |
| Governor's Energy Office | http://www.fullname of state.gov/energy/ |

*Figure 36 - RESOURCES FOR ZEC GOVERNANCE PLANNING*

## Step 8 of 10

The eighth step deals with the implementation of the energy performance renovations for existing and new buildings. A ZEC can be planned to include a combination of commercial and residential buildings, renovation of existing buildings, and/or design and construction of new buildings. The objectives, activities, and deliverables specified in this step are organized according to the type of buildings to be addressed so that ZEC planners can skip any instructions that do not apply to their project.

This step does not specifically address renovation and construction related to community infrastructure elements such as roads, parking lots, sidewalks, streetlights, water, storm water, or electrical and communication utilities, though the same general principles apply. The ZEC Project Management planning team will need to draw on the representatives of the other five of the planning-track teams, regulation, marketing, financial, technology and operations, and lead them in this process of planning and executing the requirements of this step.

Climate and siting influence a building's ability to efficiently provide a human "comfort-zone," which considers temperature, humidity, air movement, sunshine levels, and human activity. This efficiency in building performance is directly proportional to the building's energy requirement, and buildings can be designed and constructed, or renovated, to optimize the balance between climatic and site conditions with the building's energy requirements.

Conservation of energy is mainly manifest in buildings through insulating characteristics, generally referred to as the "building envelope." The performance of the building envelope results from the thermal insulating qualities and moisture barriers created by exterior walls, roofs, doors and windows, and any opening therein. Because small air or moisture leaks have large consequences, attention to detail in sealing and caulking can have significant impact on energy performance.

> *Conservation of energy in buildings is also possible when each building's design is coordinated with solar and outside airflow conditions. Passive solar designs warm structures at specific times of day according to the seasons, and prevent solar warmth at all other times. The shape and orientation of a building can improve energy efficiency by directing exterior airflow away from the building structure. Lastly, the reflection of solar energy from surrounding structures and paving can affect the energy efficiency of a building. These principals are referred to collectively as "Sol-air" design. (Olgyay and Olgyay 1963)*

Renewable energy capability can be optimized when a building's design supports the installation of solar panels. When the shape, slope, size and orientation of roof designs align with the path of the sun, or the landscape design integrates ground-mount or canopy-mounted solar, the energy generation potential is optimized.

Another form of renewable energy that can be optimized through building design is the geothermal heat pump system. Using ground loops of piping to capture thermal energy, and often incorporating additional piping within building slabs, these systems can provide radiant building heating. Less common forms of building-integrated energy supply include integrated wind generation, skylights, and daylight harvesting and distribution systems.

Every one of the foregoing energy solutions involves construction to:

☐ Build these solutions into newly constructed buildings, or

☐ Renovate existing buildings to incorporate these types of solutions.

Any activity involving renovation or new construction of buildings has very similar requirements. Whether the buildings are a hundred years old or comprise a future vision in an architect's mind, they all need to conform to similar energy goals when located within a ZEC. These buildings can be commercial offices, retail stores, condominiums, apartments, warehouses, or attached/detached residential types.

### Common Construction and Renovation Issues for All Types of Buildings

ZEC planners may encounter situations where it will be helpful for them to understand the basic consideration of building construction renovation. The planners may be selecting new homebuilders for the community, or a commercial builder to construct multi-use type facilities within the ZEC, or lining up energy performance contractors (EPC) for the ZEC occupants to engage in energy efficiency projects involving home renovation. In any of these scenarios, the following will serve the planners as a checklist of considerations:

(1.) Safety should be the number one goal in any construction activity. This includes protecting the public, occupants and workers from accidents, explosions, electrocution, toxins, entrapment and many other possible types of risks.

(2.) Protection of property from construction activities is also an important consideration. Wear and tear on streets caused by heavy construction equipment, structural collapse, fire danger and damage to gas or water lines are significant problems. On the other end of the spectrum, a contractor installing solar panels or insulation for an existing building may damage the roof, landscape, or gutters in the process.

(3.) The ability of a construction contractor to provide insurance coverage to protect the ZEC community from claims for death, personal injury, and property damage is an important qualifier. The evidence of insurability of the contractor may indicate their good safety record in the past. ZEC planners should familiarize themselves with industry norms for "insurance certificates" and designation of "loss-payees."

(4.) Cost control is a critical concern in all construction projects. Budgets vary for each project, but it is usual to have budget constraints, and energy improvements may be less of a priority than other more basic project elements that are required to complete work and make a building suitable for occupancy.

(5.) Quality is a relative measure, and varies greatly from one building project to another. In a general sense, construction work must join materials correctly, deliver surfaces and joints that are regular, even, plumb, level, and square. The construction must withstand normal cycles of usage and stresses caused by use, as well as the forces of weather and seismic activity. Structures must be watertight and colorfast and maintainable for a designated period of time from 5 - 100 years. Environmental features are typically required, such as materials that are anti-microbial and non-toxic, non-polluting and do not emit gasses ("off-gassing").  Building code-compliant construction is usually the minimum acceptable quality.

(6.)  Speed of construction is among the most significant concerns in construction and renovation, especially when the use of the building depends on completion. Design and construction professionals use critical-path analysis to determine the most effective sequence of work to achieve acceleration of the completion schedule. Contracts are often structured with performance bonuses for schedule acceleration, and liquidated damage assessments (financial penalties) for schedule delays–provided they are not due to force majeure.

(7.)  Aesthetics of any building's exterior or interior construction, including architectural decorations, fixtures, finishes and furniture that result from the construction and renovation activities, carry significant value considerations for commercial building owners, homeowners and tenants.

(8.)  Disturbances caused by new construction or renovation projects can include street closures, blocking traffic, pedestrian barricades, and noise or dust, all of which may prevent the ZEC occupants' peaceful enjoyment, and/or the ability to utilize commercial or residential building spaces.

(9.)  Climate considerations can affect the timing of construction activities when frost prevents digging and cold or damp conditions prevent opening up of buildings or the ability to progress construction. Many building materials require very specific ambient conditions during their application: concrete, construction adhesives, coatings, paint and other materials that require specific humidity and temperature requirements which can limit the acceptable schedule for their application.

(10.) Embodied energy within building materials, caused by its manufacture, transportation or installation processes may conflict with sustainability goals for the ZEC.

(11.) Security concerns can become amplified when construction activities are occurring. Security risk may result from having unfamiliar workers within a building or community, or open doorways on construction sites, or because workers are within occupied premises performing renovations, or because properties are unoccupied during construction.

(12.) *Waste created in the process of construction and renovation is often substantial, and waste generation needs to be minimized and responsibly recycled for other uses, either down-cycled through organic decomposition, up-cycled through technical reuse of resources, or disposed of using specialized procedures to neutralize environmental impacts. (McDonough and Braungart 2002)*

(13.) Certification of design and construction processes can be achieved when properly specified. In addition to the green building and sustainability standards explained in Section 16, standards certifications are available for virtually every craft—from the testing of concrete, noise-levels, fire control, and the joinery of woodworks.

(14.) The provision of warrantees for work and materials are important considerations in comparing builders.

*Objectives for Energy-Oriented Renovation and Construction Activities*

(1.)  Promote the enhancement of energy efficiency for the ZEC.

(2.)  Improve the capability of the ZEC to meet energy demands with onsite and offsite renewable energy generation.

(3.)  Build a shared vision and knowledge regarding the goals and science related to achieving the above.

(4.)  Assist the ZEC occupants in identifying resources to achieve construction and renovation goals.

(5.)  Educate the community in the requirements for professional and do-it-yourself energy projects.

(6.)  Demonstrate resources, technologies, services, and techniques that serve the ZEC's energy goals.

(7.)  Provide plan reviews and advisory support related to energy advancement for home and business owners in the ZEC.

(8.)  Provide business planning and advisory support to the community.  Help with matters related to energy advancement projects, educating the community on advantageous contract structures, financing resources, and dispute resolution alternatives.

(9.)  Provide permitting, inspection, and certification to ZEC occupants that are consistent with the ZEC role.

(10.)  Determine how the ZEC can best work with other community agencies and resources to deliver the above—and articulate the extent of the ZEC's role and responsibility with the former.

(11.)  Incorporate new team members, advisors, and service providers on the planning team.

(12.)  Review the Design Development Document.

(13.)  Determine the cost implications for construction and renovation activities.

(14.)  Determine if any additional analysis is required and how that will occur.

(15.)  Provide further validation of the project timeline.

(16.)  Review Construction Documents for planned projects.

## Notes

*Achieve the Following in This Step*

(1.)  Deliverables: Community guidebook, contact lists, sample documents, sample project schedules. Information indicating the overall scope of construction, economic impact, responsibilities of the ZEC and other agencies/resources available to the community, along with an overall schedule.

(2.)  Meetings: Leadership, team development, planning-track teams, planning workshops, vendor presentations, community meetings.

(3.)  Other activities: Team building, vendor qualification, contracting requirements (safety, quality, insurance, security, etc.)

(4.)  Outcomes: Team and ZEC project development, learning about the construction process, establishing expectations among community and service providers.

(5.)  Duration: 18-60 months

(6.)  See other Sections: Financial Considerations – Zeconomics (Section 3), Conservation (Section 6), Exploring Renewable Energy (Section 9), Other Programs and Standards (Section 14), Competitive Dialogue Process (Section 22).


*Key Questions Related to Energy-Oriented Renovation and Construction Activities*

(1.)  What is involved in constructing energy-efficient buildings that incorporate renewable energy?

(2.)  What is involved in energy-efficient retrofits to existing homes and buildings, and how can renewable energy be incorporated in those buildings?

(3.)  What is the general scope of the work required for new buildings (quantity, type, quality, size, schedule)?

(4.)  What is the general scope of the work required for the renovation of buildings? (quantity, type, quality, size, schedule)

(5.)  What developers, general contractors (GCs) homebuilders, energy performance contractors (EPCs) operate in the region, or would perhaps operate in the region given the opportunity? (NOTE:  This step is meant to be an informal introduction, not to be construed as the procurement through a Competitive Dialogue Process described separately.)

(6.)  What level of force is appropriate to support the construction and renovation for the ZEC?

*Activities for Energy-Oriented Renovation and Construction*

*Organize the following activities for the team:*
(1.)  As a group, discuss program requirements for energy efficiency and renewable energy (EERE).
(2.)  Develop a summary of ZEC construction and renovation requirements to understand and communicate the overall scope of work. (NOTE: reference Design Development Document and see planning template below this section.)
(3.)  Perform research to identify and profile developers and green builders of commercial buildings.
(4.)  Perform research to identify and profile architects, general contractors (GCs) and energy performance contractors (EPC) with experience in energy efficiency renovations for commercial, community, and cultural buildings.
(5.)  Identify whether there are special concerns related to historic buildings or cultural heritage sites within the ZEC. If there are, identify architectural specialists in preservation and renovation for those types of projects.
(6.)  Perform research to identify architects and homebuilders who provide energy-efficient residential products.
(7.)  Perform research to identify architects, general contractors and energy performance contractors experienced in energy efficiency renovation for residential buildings/houses.
(8.)  Perform research regarding building product suppliers capable of supporting do-it-yourself energy renovation projects by homeowners.
(9.)  Make site visits to see similar building construction and renovation projects.
(10.) Review trends in green building design construction and cost.
(11.) Develop an Executive Summary Report addressing the construction and renovation requirements.
(12.) Temporarily dissolve the track teams and appoint team leaders for the following construction sub-teams (as applicable):
A. New building construction (commercial, community and cultural)
B. Old building renovation (commercial, community and cultural)
C. New building construction (residential buildings)
D. Old building renovation (residential buildings)
E. Old building renovation and preservation (historical or cultural heritage)
(13.) Have each of the teams analyze and provide a report of the scope of work and the key program recommendations.
(14.) Have each of the three create a recommendation for the ZEC's energy conservation.
(15.) Determine the community's total requirement for construction and renovation.
(16.) Estimate the economic impact and job creation for achieving the community's total requirement for construction and renovation.
(17.) Draft an outline for a Construction and Renovation report.

(18.) Deliverables related to energy-oriented renovation and construction activities for
the planned ZEC:

A. Provide an Executive Summary Report that explains the general scope of requirements
and preliminary plan to execute the energy-oriented renovation and construction
required for the ZEC.

B. Incorporate all of the following in a Construction and Renovation Report:
i.  Description of the overall plan for building renovation.
ii. Description of the overall plan for new buildings.

C. Address the key elements of the plan, categorizing each group of buildings (options
typically include: existing, planned, commercial, residential and historic).

D. Determine which would be professional versus do-it-yourself projects.

E. Indicate the approximate timeline for execution of each element of the project.

F. Include an order-of-magnitude cost estimate (range) for the construction and
renovation, broken down by project elements.

G. Outline plans for programs, education, and communication outreach related to
energy conservation and renewable energy building projects.

H. Determine economic impact of Construction and Renovation projects for the ZEC,
including job creation.

***Resources***

| REQUIREMENTS | QUANTITY | AREA (SQ. FT.) | AGE (YEARS) | CONSTRUCTION (WOOD, BRICK, METAL) | CONDITION |
|---|---|---|---|---|---|
| Existing Commercial Buildings | | | | | |
| Planned Commercial Buildings | | | | | N/A |
| Existing Community Buildings: (Schools, Worship, Recreational) | | | | | |
| Panned Community Buildings: (Schools, Worship, Recreational) | | | | | N/A |
| Existing Residential | | | | | |
| Planned Residential | | | | | N/A |

*Figure 37 - SPREADSHEET TEMPLATE TO SUMMARIZE PROJECT REQUIREMENTS
ACCORDING TO USES, BUILDING TYPES, SIZE, AGE, AND CONDITION*

# DEPLOY RENEWABLE ENERGY

## *Step 9 of 10*

In the ninth step of the Zero Energy Community planning, the team will work to deploy the Renewable Energy Plan that was determined in Step 4 and leverage the Governance Plan established in Step 7. There are many parties who could be involved in the deployment of renewable energy, and in this step, the ZEC planning team shall determine who will be involved and their respective roles and responsibilities related to this deployment of renewable energy.

The deployment of renewable energy is likely to be a phased process, incrementally adding to the ZEC's energy supply from a number of sources:

(1.) Onsite equipment may include a mixture of solar, wind, hydroelectric, geothermal, biofuel, or other supply systems located on lots, building and parking canopies, and may be ground-mounted or roof-mounted.

(2.) Offsite energy may come from nearby Brownfield sites that are directly connected through overhead or underground transmission lines, or come from disconnected sites that feed power using a virtual net-meter, or be power purchased from a distant power provider and "wheeled" across the grid.

The ZEC entity may have direct responsibility or a supportive role in the deployment, or a combination of both. The ZEC entity may have to enter into business agreements on behalf of the community, and/or assume responsibilities for those contractual arrangements. To facilitate power purchases or energy services, the ZEC may have to aggregate the communities' real estate collateral into a single credit facility, or serve as a billing entity that apportions credits and charges for energy among individual property owners (as is the case if virtual net-metering is utilized by the ZEC).

The ZEC planning team will require the assistance of legal, financial, architectural, engineering and project management advisors during this step if they do not have those skills and experience on their team.

This ZEC planning step should embrace "the long view" in the planning and deployment of renewable energy. By considering the "future-watch" established at the outset of their planning process, the ZEC team can identify the issues that may arise five, ten, twenty or more years in the future and better protect their community from obsolescence, maintenance headaches, difficulty in upgrading technology, or limitations in the ability to serve the growth needs of their community.

Through this process, the ZEC team should embrace the idea that the "great is often the enemy of the good." In my experience with advanced technology deployment, being decisive, acting on the information that is available now, with the technology and resources actually available now, can produce a result that is superior to an indefinite pursuit of perfect results.

Although the task may seem daunting at first, the ZEC team should not become overwhelmed. By prioritizing the needs of the community and breaking the deployment into steps, the effort will move forward, and with each result accomplished, the remaining tasks will seem easier. As in other steps of implementing a ZEC, the planning team should strive to be inclusive, letting the community have a strong voice and a sense of self-control over their own destiny.

### *Deployment of Renewable Energy*

(1.) Determine any changes or refinements in the Renewable Energy Plan created in Step 4.

(2.) Clarify the scope of requirements for renewable energy deployment. Explain each method of energy supply, the provider(s), and the phases of the deployment.

(3.) Determine the proportions of onsite and offsite renewable energy supplies and if there will be a deficit in supplying the annual energy needs for the ZEC.

(4.) If there will be a deficit in the supply of renewable energy, based on the above, firm a plan to purchase renewable energy credits (RECs), as required, to reach net-zero.

(5.) Identify all planned deployments for each type of renewable energy.

(6.) Identify all the parties who have a role in providing renewable energy for the ZEC:
  A. Land Developer (master developer or horizontal developer)
  B. Energy service company (ESCO)
  C. Independent power producer (IPP)
  D. Utility company
  E. Commercial building owners
  F. Institutional building owners (e.g., school, hospital, government)
  G. Residential occupants

(7.) Determine the specific roles and responsibilities of each of the parties (listed above) who will be providing renewable energy for the ZEC.

(8.) Determine what the ZEC's responsibilities for the deployment entail:
  A. Direct responsibility to deploy a solution through internal resources (e.g., constructing and operating a district geothermal pipe loop, or onsite wind turbines, etc.)
  B. Direct responsibility to deploy a solution through external resources (e.g., contracting with an ESCO, IPP or utility company to provide energy services).
  C. Indirect responsibility to assist others as they deploy solutions through their own resources (e.g., assisting a utility in managing aggregated billing solutions for virtual net-metered electricity or assisting homeowners with privately owned installations). Examples of opportunities for value-creating services:
      i. Providing planning services
      ii. Organizing financing programs
      iii. Administering financial incentive programs
      iv. Providing inspections, permits, collecting fees
      v. Pre-qualifying equipment providers and/or installers
      vi. Any combination of the above responsibilities

(9.) Engage all required consultants and advisors:
  A. Legal counsel expert in energy law and contracting
  B. Accountant to advise on project finance, billing methods and cost accounting
  C. Consultants to address business or technical matters
  D. Project management resources
  E. Architects
  F. Engineers

(10.) Prioritize the plan elements, determine and describe any options in the deployment that are to be considered.

(11.) Assure that the deployment of the plan will achieve:

    A. Reliable operation through a defined life-cycle

    B. Safety in all conditions, including resistance to storms, seismic events, and flooding

    C. Long-term maintainability of equipment and grounds

    D. Access for firefighting and power turn-off (PTO) and smoke-venting capabilities for all roofs

    E. Compliance with all laws, codes, and community covenants

(12.) See other Sections:  Financial Considerations - Zeconomics (Section 3), Policy (Section 8), Exploring Renewable Energy (Section 9), Renewable Energy Plan (Section 21), Renovation and Construction (Section 25).

Notes

*Key Questions about Deployment of Renewable Energy*

(1.) What renewable energy resources were included in the Renewable Energy Plan developed during Step 9? (Section 21)

(2.) What renewable energy equipment already exists and is available for the ZEC?

(3.) What new renewable energy supplies will be deployed?

(4.) Who will be the parties responsible for the deployment of the new
Renewable Energy Plan?

    A. The ZEC governance entity

    B. The horizontal developer/land developer of a Greenfield ZEC

    C. A utility company

    D. An energy service company (ESCO)

    E. Commercial building owners

    F. Institutional building owners

    G. Homeowners

    H. Others

(5.) What are the ZEC's responsibilities regarding the deployment of the above?

(6.) Will the energy supplies procured be electrical or thermal, or both?

(7.) What services does the ZEC intend to utilize?

    A. Independent power producer(s) (IPPs) to supply renewable energy over the grid?

    B. A utility company to supply renewable energy or provide for the "wheeling" of offsite power from IPPs?

    C. Will the ZEC contract for energy services with an ESCO to develop and operate renewable
energy generation assets for the ZEC?

(8.) Will the ZEC develop a private power line between the ZEC and an offsite power generation facility?

(9.) Will this include development of a right-of-way?

(10.) Will this include constructing overhead or underground transmission lines, or piping (in the case of thermal energy)?

(11.) Where will the deployed renewable energy equipment be located?

    A. Within the ZEC common areas

B. Upon a remote land site

C. On the property of a utility company

D. On an energy service company's (ESCO) onsite or offsite property

E. On commercial buildings

F. On homes

G. Elsewhere

(12.) Will the ZEC contract with an installer to install equipment on common areas of the ZEC?

(13.) What professional services will the ZEC entity require?

A. Legal advisor with energy regulatory experience

B. Accountants

C. Technical consultant

D. Business consultant

E. Engineers

F. Architects

G. Project managers

(14.) What are the logical phases for the deployment considering that:

A. An onsite biofuel plant may require a long lead-time to engineer, procure, and deploy.

B. A district geothermal pipe-loop deployment may be better accomplished before street improvements are planned and made.

C. Solar canopies at commercial buildings need to be deployed in sequence with the building development.

D. Homeowners' deployments of solar may be sequenced in several batches in order to manage a smoother process.

E. The cost of drilling geothermal wells may be significantly reduced when the equipment and workers can move quickly from one property to the next.

F. What is the timeline for the deployment(s)?

(15.) What is the source of the funding for the costs that the ZEC entity is responsible for paying?

(16.) What credit issues need to be addressed on the part of the ZEC?

(17.) How will the following considerations of long-term energy deployments be addressed:

A. Maintainability - Technology that can provide reliable service during maintenance cycles is described as having "fully maintainable reliability." Is this required?

B. Upgrade-ability - What are the likely upgrades to be made in the future, and how are those upgrade pathways engineered into the initial deployment?

C. Expansion capability - What is the planned growth of the ZEC and how might the deployment need to be expanded? Considering requirements for space, interface and physical support in the engineering of the initial deployment may provide manifold benefits in the future—is this required?

D. Warrantee fulfillment responsibilities may become material issues in the future. How is the performance of service addressed in the deployment so that routine and emergency service is practical and affordable in the future?

(18.) Activities for Deployment of Renewable Energy at the ZEC:

A. Clarify the scope of requirements by reviewing the Renewable Energy Plan.

B. Briefings to advisors and consultants regarding the needs of the ZEC and the ZEC deployment plan.

C. Determine the business plan for the ZEC entity.

D. Develop a financial model for all sources and uses of funds.

E. Consider the logical sequence for the renewable energy deployment and develop a schedule based on resource availability and prerequisites for each deployment. Use a critical-path resource scheduling software to develop a project schedule (e.g., Microsoft Project, Primavera).

F. Develop scenarios for each aspect of deployment and discuss them with those affected by, and/or supporting, that deployment.

G. Engage the community in the consideration of options and decision-making.

(19.) Deliverables related to the deployment of renewable energy:

A. An executive summary describing the goals, phases and elements of the renewable energy deployment, and the financial impacts and respective benefits.

B. A schedule that indicates all major elements and phases of the planned deployment of renewable energy.

C. A financial plan that includes historical and projected sources and uses of funds, cash flow, income, and balances.

D. A document describing the credit facilities of the ZEC entity, explaining owned assets and enumerating the underlying ability to attach liens to any ZEC property owned by a third party.

E. A document containing legal opinions regarding the ZEC entity's rights concerning energy transactions.

F. Contractual agreements, including power purchase agreements (PPA) between the ZEC entity and third-party energy suppliers, energy transmission providers, energy service companies, equipment suppliers, service providers, software providers as well as fuel, and material suppliers.

G. Memorandum(s) of Understanding, laying out the general intentions and understanding of each party regarding an agreement, including ownership of renewable energy credits (RECs).

H. Terms Sheet(s) prescribing all terms and conditions, each parties' responsibilities, use and venue of law, and other requirements as advised by ZEC's legal counsel.

I. Definitive Agreement(s), including binding contracts prescribing all terms and conditions, each parties' responsibilities, rights and remedies, use and venue of law, and other requirements as advised by ZEC's legal counsel.

(20.) Resources:

Primavera: http://www.oracle.com/us/products/applications/primavera/overview/index.html

Microsoft Project: http://office.microsoft.com/en-us/project/

## Notes

# MEASUREMENT, EDUCATION AND OUTREACH COMMUNICATIONS

## *Step 10 of 10*

The tenth and last step of initiating a Zero Energy Community (ZEC) addresses the requirements that span the entire life of a ZEC project, from inception through its future. Born in response to the near and present problems of energy, environment, social equity and economics, the ZEC's construct provides a vehicle for change.

The ZEC promises to provide a path for relocating the source of energy required for transportation and buildings from fossil fuels to renewable energy. By reducing consumption through conservation, and re-powering our buildings and cars with renewable energy, a ZEC accrues manifold benefits for the environment and economy, plus it offers improved health and quality-of-life for current and future generations.

This step is the last of the planning stages and marks a transition of responsibility for the ZEC. The leaders, initiators, planners, and developers who created the project can pass on responsibility to a new and emerging ZEC community, its occupants, and leadership. Utilizing accounting, marketing, real estate, and communication professionals is essential to the successful outcome of this step.

The transition can provide a great forum for celebration of accomplishment, acknowledgement, and opportunity for new leaders; public recognition of the significant efforts and contributions of the ZEC's founders is an important part of such a transition. The transition also provides an opportunity to convey the history and institutional knowledge developed in the process of creating the ZEC to constituents who become the new owners of the vision, mission, and values.

Bring the founding team and the new participants together during this step to discuss the measurement of accomplishments, educate new participants about the ZEC features and benefits, and convey the founders' intents and aspirations for the ZEC. This line of discussion will support the group's overall goals in determining the appropriate metrics and education programs, and it will help form effective messages for external communication going forward.

The process of planning and developing the ZEC involves many considerations and planning processes, as well as the adoption of new behaviors. The overall result, however, is quite simple, and the measurement, education, and communication outreach must be bold and simple.

The simplest definition of a ZEC is "a connected community that depends on renewable energy to power its electric cars, homes, and appliances." Those who create ZECs are the pioneers of the future of energy. Leaders who develop ZECs can improve the quality of life for everybody within a community so that people can enjoy life and health, improve real estate values, and build a sense of pride among those working or living within the ZEC.

Advertising the community's accomplishment to the world, backed with credible proof of performance, will create value for the ZEC—and better the world.

### Measurement of a ZEC's Performance

Measurement of your ZEC's performance can substantiate claims of success with qualitative, (Creswell 2007) and quantitative (Starbird 2006) statistical and other data illustrating the ZEC development's benefits to individuals, businesses, the community, and nation. Measurements can serve to dislodge limiting beliefs about ZECs and make the case for investment in the ZEC model. Apply rigor in your research and make comprehensive evaluations that are objective and measurable; the resultant research reports need to support your claims.

While empirical evidence of accomplishments is necessary, do not be concerned about conveying the human elements of the ZEC's brand, as well. An individual's own story about how the ZEC affected them personally can provide compelling equity to the ZEC brand, and may actually connect better with some audiences.

Revisit your existing communications to see if the messages or audiences have changed. Determine the effect you want to achieve with those audiences and what compelling arguments you will make to cause that effect—whether the messages are promotional advertising or for internal programs to increase use of clotheslines, to promote electric-cars, or to explain high energy-efficiency media rooms, the proper use of measurements, education, and communications outreach can all help the ZEC achieve a net-zero energy balance.

Use baselines to contrast ZEC and non-ZEC communities. Focus on the audiences you want to influence and the metrics that authenticate your messages.

### ZEC Education

Education about energy use and conservation can help members of the ZEC community optimize the ZEC's performance and their own energy cost. Educating the broader community can create awareness and interest in the environmental and economic benefits provided by ZECs, creating demand, and thereby improving the value of ZEC property. Education can also explain important differentiations between authentic ZECs and inferior imitations.

### Outreach

The ZEC's communication program can extend to all of the ZEC's target audiences. Sharing information about the ZEC teams' accomplishments and the project's features, benefits, and needs may provide valuable advertising, promotion, and customer acquisition. Outreach can connect the ZEC with other collaborators whose missions align with the ZEC's needs.

### Measurement, Education, and Outreach Communications Objectives

(1.) Desired outcomes of Step 10:

    A. An aspiring community will, shared vision, and commitment across the ZEC community

    B. *Productive dialogue amongst the ZEC founders and the current leadership team (Senge 2009)*

    C. Reference points for comparison of the ZEC with the status-quo (non-ZEC)

    D. ZEC performance metrics, using the triple-bottom-line (TBL) analysis method, that will provide a meaningful indication of the ZEC's performance

    E. A plan for the frequency of reporting the progress of the ZEC

    F. Certifications for the appropriate ZEC accomplishments (e.g., ecocity, neighborhood design, green building, zero energy community and website)

    G. An understanding of how well the ZEC has measured up against the original goal and how those goals might need to change in the future

(2.) Determine opportunities and requirements for the ZEC's outreach communications.

(3.) Identify and target constituents through media channels such as broadcast television, press, social media, symposia, and other events.

(4.) Provide ZEC leadership, staff, and community with information regarding opportunities and problems.

(5.) Create an understanding of the data requirements and record keeping appropriate to support increasingly valuable feedback on ZEC performance.

(6.) Continue collaboration by seeking to optimize, innovate, and celebrate achievement.

(7.) Acknowledge the organizations, groups and individuals who contributed to the ZEC's success.

(8.) Create brand messages appropriate to each audience of the ZEC.

### Other Outcomes of Step 10

(1.) Deliverables: Final Design Development Document

(2.) Meetings: ZEC Leadership, community development, community workshops, original planning-track team meetings, certification team meetings, and celebrations for achievement of goals

(3.) Other activities: Core team building, business intelligence, and ongoing research and evaluation

(4.) Outcomes: Team development, measurement, education and outreach communications

(5.) Duration: 45-60 days for planning, then perpetual program

(6.) See other sections: Living and Working in a ZEC (Section 2), Financial Considerations–Zeconomics (Section 3), Electric Vehicles (Section 10), Other Programs and Standards (Section 14), Certification of a ZEC (Section 16), Communications for a ZEC (Section 17)

*Figure 38 – TESTIMONIALS DEMONSTRATE THE SOFT VIRTUES OF A ZEC*

"Living at the Hillstation ZEC has given our family a common purpose and made us feel a part of this community. The money we save on electricity, natural gas and gasoline is going into our children's college fund." (Sample)

Notes

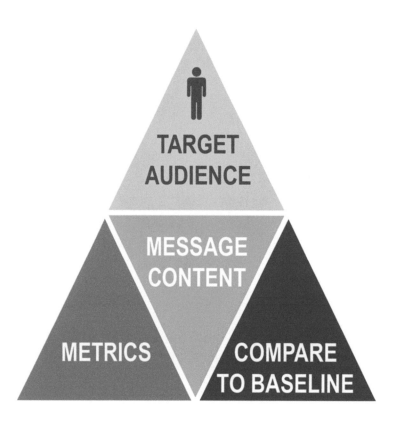

*Figure 39 - PURPOSEFUL COMMUNICATIONS THAT PRODUCE AN INTENDED EFFECT BECAUSE THEY ARE SPECIFIC TO EACH AUDIENCE GROUP. DISTINCT MESSAGES RELATE CHANGE FROM A BASELINE AND METRICS FOR COMPARING THAT CHANGE.*

### Measurement, Education, and Outreach Communications Questions

(1.)  Who among the ZEC founders and current leadership and team should be involved in this step?  How can they be encouraged to shed their roles and participate as individuals working together towards a shared vision and group commitment? (Senge 2009)

(2.)  How can the ZEC team measure and respond to the original goals of the ZEC?

(3.)  What are the ZEC's goals for improvement and how might the metrics indicating system performance and human behaviors be used to attain those goals?

(4.)  Relative to future goals for the ZEC, who are the groups that may react/respond to measurements?  See examples below:

    A. Businesses and residents may react to metrics that inform them about how to reduce energy costs.

    B. Real estate professionals, including brokers and appraisers, may respond to information regarding transactions, appreciation, and cost of ownership.

    C. Merchants may respond to information regarding pedestrian traffic and demographics.

    D. Commercial building developers and homebuilders may respond to information about pre-sales, business cycles, and volumes.

> E. The general public may respond to information about the overall reduction in greenhouse gas emissions and carbon consumption.
>
> F. Occupants may respond to current energy performance information by reducing or deferring energy use—perhaps by applications and messaging their smart phones or using other prompts.
>
> G. Auto buyers may respond to information regarding fuel savings and pollution reduction realized by ZEC occupants who drive electric-vehicles.
>
> H. ZEC occupants may proactively conserve energy if they have information indicating the imminent changes in weather.

(5.) How can the ZEC develop baseline measurements that provide points for comparison of actual and projected conditions that show how the ZEC is performing?

(6.) What metrics provide the most meaningful indication of how well the ZEC's energy resources are being managed?

(7.) What metrics provide the most meaningful indication of the ZEC's financial performance?

(8.) What comparative measurements would be most convincing and affect behavior change regarding energy use?

(9.) What benchmarks are important to the ZEC's operation and comparison with other ZECs?

(10.) What metrics should be combined into more meaningful data sets, or "metadata"?

(11.) How can the ZEC provide a feedback mechanism for continual improvement?

(12.) How can the ZEC promote continuum of energy conservation, addressing the following:

> A. Lifestyle-related energy efficiency in home, office or retail building:
>> i.   Temperature requirements
>> ii.  Turning off unused lights and appliances
>> iii. Time-of-use relative to renewable energy production and demand
>> iv.  Closing doors and windows when efficient
>
> B. Transportation-related energy:
>> i.  Use of car-share, car-pooling and mass-transit
>> ii. Adoption of renewable energy powered electric vehicles (EVs)
>
> C. Energy monitoring, control and efficiency technologies:
>> i.   Real-time monitoring of energy generation, use and storage
>> ii.  Automation of systems for efficiency
>> iii. Use of apps to provide efficiency alternatives
>> iv.  Sharing EV batteries with ZEC buildings

*Measurement, Education, and Outreach Communications Activities*

(1.) Organize a workshop-style meeting for the founders and leaders of the ZEC to collaborate about measurement, education, and outreach communications:

A. Ask the group about the outcomes it sees as being possible

B. What does that group have in terms of assets?

C. How can they envision the adoption of (more) ZEC sites at their place—town, campus or neighborhood?

(2.) Conduct a round-table discussion and create a preliminary list of:

A. Measurement goals

B. Education goals

C. Communication outreach goals

(3.) Invite an organizational development expert, facilitator, behavioral psychology researcher, accountant, and a statistician to participate in the workshop.

(4.) Involve an accounting firm in developing and applying a measurement system for the ZEC. Because sustainable accounting systems are a burgeoning area of practice, many firms may seize the opportunity and even provide pro bono services. The accounting firm's name brand may improve the credibility of the ZEC's claims of success.

(5.) Revisit the audiences defined in earlier steps of the ZEC creation and communications activities, and adjust appropriately.

(6.) Discuss the types of effects the ZEC seeks in order to meet its goals for success.

(7.) Discuss the appropriateness of other standards and certification programs.

(8.) Develop an understanding of the types of measurements and specific metrics that are not outputs from the preferred standards and certifications.

(9.) Determine what educational components are required to achieve the positive effects that improve measurements.

(10.) Discuss methodologies for educationally-oriented programs for community, youth, student groups, families, businesses, merchants, and builders—those that are both internal and external to the ZEC.

(11.) Identify community organizations such as schools, scouts, clubs and business chambers that can collaborate to deliver education.

(12.) Consider many possible elements of the ZEC and its brand. Edit the list and prioritize.

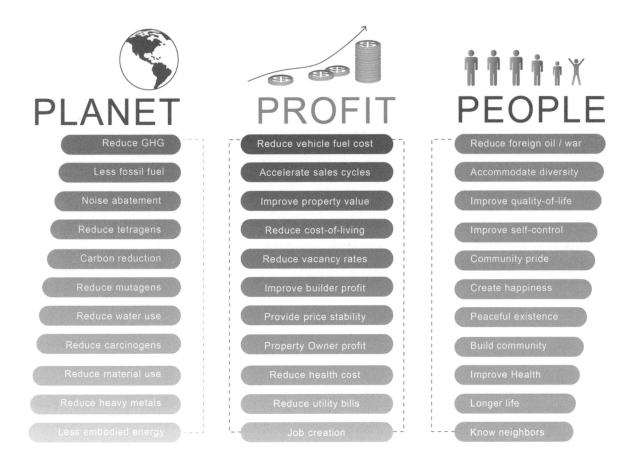

**PLANET**

- Reduce GHG
- Less fossil fuel
- Noise abatement
- Reduce tetragens
- Carbon reduction
- Reduce mutagens
- Reduce water use
- Reduce carcinogens
- Reduce material use
- Reduce heavy metals
- Less embodied energy

**PROFIT**

- Reduce vehicle fuel cost
- Accelerate sales cycles
- Improve property value
- Reduce cost-of-living
- Reduce vacancy rates
- Improve builder profit
- Provide price stability
- Property Owner profit
- Reduce health cost
- Reduce utility bills
- Job creation

**PEOPLE**

- Reduce foreign oil / war
- Accommodate diversity
- Improve quality-of-life
- Improve self-control
- Community pride
- Create happiness
- Peaceful existence
- Build community
- Improve Health
- Longer life
- Know neighbors

*Figure 40 - TRIPLE BOTTOM LINE ILLUSTRATION*

(13.) Develop comparisons of ZEC performance with other communities that have not
implemented ZECs:
   A. Learn about their energy and building practices and statistics
   B. Study their real estate values, rents, and price trends
   C. Examine data about energy use and related emissions
   D. Make adjustments based on variances in the sites being compared
   E. Make comparisons that demonstrate improvements by the ZEC over the averages
(14.) Determine what metrics will be regularly reported and the frequency of reporting (e.g.,
using a time frame of real-time, monthly, quarterly, annually, life, or since inception).
(15.) Determine which platforms the data will be reported on (e.g., television, print, digital
signage, website, blog, tweet, word-of-mouth, and whether it is
confidential/not reported).
(16.) Determine which apps can be deployed or created to advance the measurements,
education and communications outreach.
(17.) Consider sponsorship alignments to meet ZEC goals. Would it be worthwhile to have a

homebuilder or a manufacturer of solar panels, windows or electric cars sponsor a promotion for your ZEC?  Consider the bi-lateral impacts of such partnerships.

(18.) What other activities serve the objectives of the measurements, education and communications outreach process (e.g., public relations, collective effort with other ZECs, community activities, sponsorships, media relations, etc.)?

(19.) Organize "ribbon-cutting" events.

### Measurement, Education, and Outreach Communications Deliverables

(1.)  Provide a report on the plans for measurements, education, and communications outreach. (This may be several reports that address individual aspects and an overall executive summary.)

(2.)  Initiate internal communications, advertising and public relations to promote the adoption of the above plan(s).

(3.)  Create metrics that provide feedback regarding the performance of the plans for measurements, education, and communications outreach.

(4.)  Create a celebration—congratulations are in order!

Notes

*Figure 41 - THE TALLEST BUILDING IN CHINA IS SHANGHAI TOWER, A MASSIVE GREEN BUILDING THAT TOPPED OUT IN 2013. GENSLER ARCHITECTS AND DESIGNERS, DESIGNED IT TO WORK WITH THE ENVIRONMENT.*

*LIST OF COMMUNITY ENERGY PROJECT WEBSITES*

| DEVELOPMENT | WEBSITE |
|---|---|
| Clean Energy Collective | http://nwcommunityenergy.org/solar/solar-case-studies/CEC_colorado |
| Cooperative Solar Farm | http://www.unitedpower.com/mainNav/greenPower/solPartners.aspx |
| Dunedin | http://www.mnn.com/your-home/remodeling-design/blogs/net-zero-housing-done-right-in-dunedin |
| FortZED | http://fortzed.com/ |
| GEOS Community | http://discovergeos.com/ |
| Green Acres | http://www.greenacresnewpaltz.com/ |
| Hoa'âina | https://www1.eere.energy.gov/office_eere/pdfs/webinar_hawaii_kaupuni.pdf |
| Jetson Green | http://www.jetsongreen.com/2010/01/net-zero-energy-paradigm-pilot-project-lafayette.html |
| Lowry - Boulevard One | http://lowryredevelopment.org/annex/ |
|  | http://lowryredevelopment.org/annex/wp-content/uploads/2012/12/GDP-public-mtg-12-10-12-for-webpptx.pdf |
| Solar for Sakai | http://nwcommunityenergy.org/solar/solar-case-studies/copy2_of_the-vineyard-energy-project |
| University Park Solar | http://universityparksolar.com/ |
| West Village | http://westvillage.ucdavis.edu/ |
|  | https://www1.eere.energy.gov/office_eere/pdfs/webinar_ucdavid_west_village.pdf |
| zHome | http://www.dwell.com/renovation/article/zero-energy-community-final-post |
|  | http://www.dwell.com/renovation/article/zero-energy-community-part-10 |
|  | http://www.dwell.com/renovation/article/building-zero-energy-community-part-7 |
|  | http://www.dwell.com/renovation/article/zero-energy-community-part-6 |
|  | http://www.dwell.com/renovation/article/zero-energy-community-part-1 |
|  | http://www.dwell.com/design-101/article/zeroing |

*Figure 42 - LIST OF COMMUNITY ENERGY PROJECT WEBSITES*

## ABBREVIATIONS

There are many unique terms associated with the Zero Energy Community. This section provides the definitions for all of the abbreviations that are used in the guide listed within the categories listed below:

a) Agencies
b) Businesses
c) Development Tools
d) Energy Technology
e) Financial
f) General
g) Laboratories
h) Policy
i) Procurement
j) Rating
k) Scientific
l) Territory
m) Transportation

### Agencies

BBC British Broadcasting Company
CCD City and County of Denver
CDC Centers for Disease Control
CERT Committee on Energy Research and Technology
DOD Department of Defense
DOE Department of Energy
DOT Department of Transportation
EPA Environmental Protection Agency
EU European Union
FERC Federal Energy Reliability Corporation
GEO Governor's Energy Office
GSA Government Services Administration
HOA Homeowner's Association
IEA International Energy Agency
ILFI International Living Future Institute
ISI Institute for Sustainable Infrastructure
NAHB National Association of Home Builders
NIH National Institute of Health
PHI Passive House Institute
PUC Public Utility Commission
SHOA Sustainable Homeowner's Association

TZO TowardZero.org
UNFCCC United Nations Framework Convention on Climate Change
USGBCUS Green Building Council
USGS United States Geological Service
WCED United Nations World Commission on Environment and Development
WHO World Health Organization

### Businesses

GE General Electric Company
EPC Energy Performance Contractor
ESCO Energy Services Company
GC General Contractor
IOU Investor Owned Utility
IPP Independent Power Producer
TMC Toyota Motor Corporation

### Development Tools

CPM Critical Path Method
GDP General Development Plan

### Energy Technology

| | |
|---|---|
| AC | Alternating Current |
| ADG | Anaerobic Digester Gas |
| BMS | Building Management System |
| BTU | British Thermal Unit |
| CHP | Combined Heat and Power |
| DC | Direct Current |
| DEG | Distributed Energy Generation |
| DG | Distributed Generation |
| DSM | Demand-side Management |
| EE | Energy Efficiency |
| EERE | Energy Efficiency and Renewable |

### Energy

| | |
|---|---|
| EIA | Energy Information Administration |
| EMS | Electrical Management System |
| GHP | Geothermal Heat Pump |
| HAWT | Horizontal Axis Wind Turbine |
| HVAC | Heating Ventilation Air Conditioning |
| Hz | Hertz |
| iBMS | Intelligent Building Management System |
| kW | Kilowatt (one thousand) |

### Financial

| | |
|---|---|
| ADB | Asian Development Bank |
| BLN | Billion |
| CAGR | Compound Annual Growth Rate |
| EIN | Employer Identification Number |
| MLN | Million |
| PACE | Property Assessed Clean Energy |
| PITI | Principle Interest Tax and Insurance |
| PITIU | Principle Interest Tax Insurance and Utilities |
| PPA | Power Purchase Agreement |
| REIT | Real Estate Investment Trust |
| SAVE | Sensible Accounting to Value Energy Act of 2013 (S. 1106) |
| TIF | Tax Incremental Financing |
| USD | US Dollar (Also $) |

### General

| | |
|---|---|
| [SIC] | sic erat scriptum, "thus was it written" e.g. For example Hr. / Hrs.   Hour / Hours |
| ICT | Information and Communications Technology |

### Laboratories

| | |
|---|---|
| e-Lab | Electricity Laboratory – RMI |
| NREL | National Renewable Energy Laboratory |
| RMI | Rocky Mountain Institute Policy |
| ARRA | American Reinvestment and Recovery Act of 2009 |
| EIA | Environmental Impact Analysis |
| REC | Renewable Energy Credit |
| RPS | Renewable Portfolio Standard |
| SREC | Solar Renewable Energy Credit |

### Procurement

| | |
|---|---|
| BOQ | Bill of Quantities |
| FAT | Factory Acceptance Test |
| KPI | Key Performance Indicator |
| NDA | Non-Disclosure Agreement |
| RFI | Request for Information |
| RFP | Request for Proposal |
| RFQ | Request for Qualifications |
| SAT | System Acceptance Test |
| SLA | Service Level Agreement |
| SOQ | Statement of Qualification |
| SOW | Scope of Work |

### Rating

| | |
|---|---|
| AEO | Average Energy Output of Wind |
| HERS | Household Energy Rating System |
| SEER | Seasonal Energy Efficiency Ratio |

### Scientific

C02     Carbon Dioxide
IAQ     Indoor Air Quality
N0x     Nitrogen Oxide
QUAD    Quadrillion: one-thousand-
        million-million, signified by 1015
        and the SI prefix peta-

### Sustainability

LEED    Leadership in Environmental Design
LEED ND   LEED Neighborhood Design
ND      Neighborhood Design
PHIUS   Passive House Institute – US
TBL     Triple Bottom Line (also 3BL)

### Territory

US      United States of America
USA     United States of America

### Transportation

EV      Electric Vehicle
HEV     Hybrid Electric Vehicle
ICE     Internal Combustion Engine
PEV     Plug-in Electric Vehicle
PHEV    Plug-in Hybrid Electric Vehicle

Andrews, David, interview by John Whitcomb. 2013. Project Manager, The Lowry Redevelopment Authority (April 22).

Asian Development Bank. 2009. ADB.org. October. Accessed October 2009. http://www.adb. org/.

Asmus, Peter. 2012. Small Wind Market Growth. Industry Analysis, Boulder: Pike Research.

Ball State University. 2013. Going Geothermal. August 30. Accessed August 30, 2013. http:// cms.bsu.edu/about/geothermal.

Barclay, Bernays, interview by John Whitcomb. 2013. Investment banker for power, renewable energy, infrastructure; legal counsel to entrepreneurs and project developers (April 6).

BBC 2012. Nissan LEAF advertisement aired in India. New Delhi, November 5.

Beatley, T., and K Manning. 1997. Ecology of Place: Planning for Environment, Economy, and Community. Washington D.C.: Island Press– Covelo.

Berwyn, Bob. 2012. "Large Scale Forest Biomass Energy Not Sustainable." Summit County Citizen, April 19.

Biomass Energy Centre. 2013. What is Biomass? May 24. Accessed May 24, 2013. http://www. biomassenergycentre.org.uk/portal/page?_pageid=76,15049&_dad=portal&_schema=PORTAL.

Bobrow-Williams, Sarah, interview by John Whitcomb. 2013. Goddard College, community development expert (April 23).

Boggs, Patton, L.L.P. 2009. American Recovery and Reinvestment Act of 2009 - Programmatic Analysis. Legislative Analysis, Washington D.C.: Patton Boggs L.L.P.

Bradsher, Keith. 2013. "US and Europe Prepare to Settle Chinese Solar Panel Cases." New York Times, May 20: 2013.

Bratley, James. 2007. Clean Energy Ideas. May. Accessed May 25, 2013. http://www.clean-energy-ideas.com/energy/energy-dictionary/biofuel-definition.

Breakthrough Technologies Institute. 1993. Fuelcell 2000. Accessed May 23, 2013. http://www. fuelcells.org/.

Brett, Deborah L., and Adrienne Schmitz. 2009. Real Estate Market Analysis: Methods and Case Studies, Second Edition: Washington D.C.: Urban Land Institute.

California Energy Commission. 1999. Home Energy Rating System (HERS) Program. June 17. Accessed April 16, 2012. http://www.energy.ca.gov/HERS/.

Calpine Corporation. 2012. The Geysers. Accessed April 23, 2013. http://www.geysers.com/.

Carlisle, AIA, Nancy, Otto VanGeet, and Shanti Pless. 2009. Definition of a "Zero Net Energy" Community. Technical Report - NREL/TP-7A2-46065, Golden: National Renewable Energy Laboratory.

Centers for Disease Control and Prevention. 2010. Healthy Places. July 21. Accessed June 3, 2011. http://www.cdc.gov/healthyplaces/about.htm.

Centers for Disease Control. 2013. "Public Health Terms for Planners & Planning Terms for Public Health Professionals." Atlanta, GA. http://www.cdc.gov.

Cherian, Sunil, interview by John Whitcomb. 2013. Chief Executive Officer, Spirea, Smart Grid entrepreneur (April 22).

Chiras, Daniel. 2006. The Homeowner's Guide to Renewable Energy: achieving energy independence through solar, wind, biomass and hydropower. Gabriola Island: New Society.

Civic Impulse, LLC. 2013. Govtrack.us. June 6. Accessed June 6, 2013. http://www.govtrack.us/congress/bills/113/s1106.

Committee on Science and Technology for Countering Terrorism, National Research Council of the National Academies. 2002. Making the Nation Safer – The Role of Science and Technology in Countering Terrorism. Washington D.C.: The National Academic Press.

Crandall, Kelly, interview by John Whitcomb. 2013. Sustainability Specialist, City of Builder, Colorado (May 15).

Creswell, John W. 2007. Qualitative Inquiry and Research Design: Choosing among Five Approaches -Second Edition. Thousand Oaks: Sage Publications.

Dameron, Mark. 2013. Chief Marketing Officer, EquityLock Solutions, Inc (February 12). 2008. Demand Response Symposium: industry action plan at PJM Interconnect. Paula DuPont-

Kidd. 8 13. Accessed 11 22, 2008. http://www.reuters.com/article/2008/05/16/idUS206497+16-May-2008+PRN20080516.

Demonstration, Panasonic Product. 2012. Consumer Electronics Show 2012. CES. Las Vegas: CEA.

Design Workshop. 2007. Toward Legacy. Easthampton: Spacemaker Press.

Dorsey, Judy. 2012. "Smart Grid Live - 2012." Brendle Group. Fort Collins: September.

Driskell, David, interview by John Whitcomb. 2013. Executive Director, Community Planning and Sustainability, City of Boulder, Colorado (April 24).

DuVivierr, K. K. 2011. Renewable Energy Reader. Durham: Academic Press.

Electric Drive Transportation Association. 2013. Cumulative US Plug-in Vehicle Sales. January. Accessed April 7, 2013. http://www.electricdrive.org/index.php?display=GeneralSearch&action=AddSearchTermAction&searchstring=electric+vehicle+sales.

Energy Department's Office of Energy Efficiency and Renewable Energy. 2012. Energy Department Announces New Partnership to Certify Zero Net-Energy Ready Homes. August 20. Accessed August 20, 2012. http://apps1.eere.energy.gov/news/progress_alerts.cfm/pa_id=787.

Everblue. 2011. What is Carbon Accounting? - Greening Corporate America and Wall Street. February 16. Accessed May 22, 2013. http://www.everblue.edu.

Fiona, Harvey. 2010. Green vision: the search for the ideal Eco-City. September 7. Accessed November 21, 2011. http://www.ft.com/intl/cms/s/c13677ce-b062-11df- 8c04-00144feabdc0,Authorised=false.html?_i_location=http%3A%2F%2Fwww.
ft.com%2Fcms%2Fs%2F0%2Fc13677ce-b062-11df-8c04-00144feabdc0.
html%3Fsiteedition%3Di ntl&siteedition=intl&_i_referer=http%3A%2F%2Fsearch.ft.c.

Force, Montgomery, interview by John Whitcomb. 2013. Owner of Force Consulting and Executive Director at Lowry Redevelopment Authority (February 11).

Freidman, Thomas L. 2009. Hot, Flat, and Crowded: why we need a green revolution, and how it can renew America. Release 2.0 – Updated and Expanded. New York City: Picador / Farrar, Straus and Giroux.

GeoExchange. 2013. May 15. Accessed May 15, 2013. http://GeoExchange.org. Goodman, Diane. 2011. Promoting Diversity and Social Justice. New York City: Routledge. Gore, Al. 2013. CBS Television Network - David Letterman Show. New York City, March 28.

Governor Bill Ritter, Colorado (D). 2012. "Director, Center for the New Energy Economy, Colorado State University." Keynote. Aspen: Montreux Energy, June 6.

Graedel, Thomas. 2011. Industrial Ecology and the Ecocity. Accessed November 21, 2011. http://www.nae.edu/Publications/Bridge/ UrbanizationEngineering/ IndustrialEcologyandtheEcocity.aspx.

Hammack, Katherine. 2010. Assistant Secretary of the Army (Installations, Energy & Environment). Presentation, Denver: Brownstein Hyatt Farber Schreck, LLP.

Hawken, Paul. 1993. The Ecology of Commerce. New York City: HarperCollins.

Hawkin, P., Lovins, A. & Lovins, H. 1999. Natural Capitalism. New York City: Little, Brown and Company.

Hawkin, Paul. 1993, 1999.

Hirsch, Arthur, interview by John Whitcomb. 2012. Environmental Consultant (June 21).

Horsting, Walter. 2013. "Thorium Molten Salt Reactors." Calfornia Energy Commission Workshop. Sacramento: Business Development International. 22.

Institute for Sustainable Infrastructure. 2012. Envision Rating System. Accessed June 20, 2012. http://www.sustainableinfrastructure.org/rating/.

International Living Future Institute. 2010. Living Building Challenge. Accessed June 22, 2013. http://living-future.org/.

Investopedia. 1999. Definition of Biofuel. Accessed May 24, 2012. http://www.investopedia. com/terms/b/biofuel.asp.

Investopedia. 2012. Dictionary - Principal, Interest, Taxes, Insurance - PITI. August 21. Accessed August 21, 2012. http://www.investopedia.com/terms/p/piti.asp#ixzz29Pvo5PXo.

James, Ben. 2011. "Federal Energy Regulatory Commission (FERC) Revised Power Grid Access Rule - Order 1000." LAW 360. August 11. Accessed May 7, 2013. http://www.law360. com/ articles/263377/a-full-account-of-ferc-order-1000.

Johnson, Todd, interview by John Whitcomb. 2013. Partner, Design Workshop, Inc. (February 11).

Jonathan Fahey. 2013. "US Power Grid Cost Rise, but Service Slips." Associated Press, Big Story. March. Accessed May 10, 2013. http://bigstory.ap.org/article/us-power-grid-costs-rise-service-slips.

Jones, Alex. 2013. Infowars. August 21. Accessed August 28, 2013. http://www.infowars. com/ west-coast-of-north-america-to-be-hit-hard-by-fukushima-radiation/.

Kaplan, Stan, Fred Sissine, Amy Abel, Jon Wellinghoff, Suedeen Kelly, and James Hoecker. 2009. Smart Grid Modernizing Electric Power Transmission and Distribution; Energy Independence, Storage and Security, Energy Independence and Security Act of 2007: Improving Electrical Grid Efficiency, Communication, Reliability, and Resiliency; Integration. Government Sponsored Research, Alexandria: Compiled by The Capital.Net Inc.

Katherine, I. 2006. The private finance initiative: How to conduct a competitive dialogue procedure. Leeds: United Kingdom—Department of Health Private Finance Unit.

Kawasaki, Guy, interview by John Whitcomb. 2011. Author / Entrepreneur (October 26).

Keith, John, interview by John Whitcomb. 2013. President, Harvard Communities (March 25).

Kelly-Detwiler, Peter. 2013. "Energy Storage: Continuing to Evolve." Forbes, May 13.

Kerpin, Phil. 2012. Fox News. May 22. Accessed May 15, 2013. http://www.foxnews.com/opinion/2012/05/22/obamas-war-on-coal-hits-your-electric-bill/.

Kind, Peter. 2013. "Disruptive Challenges – Financial Implications and Strategic Responses to a Changing Retail Electric Business." Edison Electric Institute.

KMW ENERGI Inc. 2013. Biomass Energy Technology - Making History... Our Future. August 28. Accessed August 28, 2013. http://www.kmwenergy.com/.

Lee, Gus. 2006. Courage – The Backbone of Integrity. Hoboken: Jossey-Bass.

2012. LEED-certified Building Stock Swells to Two-Billion Square Feet Worldwide. July 26. Accessed July 26, 2012. http://www.usgbc.org/articles/leed-certified-building-stock-swells-two-billion-square-feet-worldwide.

Leiserowitz, A., E. Maibach, C. Roser-Renouf, and J. Hmielowski. 2011. "Global Warming's Six Americas / Yale Project on Climate Change." George Mason University Center for Climate Change Communication 66.

Lester, J. 2013. "CBO Analysis Shows Benefits of a Carbon Tax, If Implemented Carefully." Clean Finance Journal. May 24. Accessed May 24, 2013. http://www.cleanfinance.net.

Lovins, Amory, interview by John Whitcomb. 2013. Chief Scientist and Founder, Rocky Mountain Institute (July 1).

—. 2002. "Innovation in Efficient Vehicle Design and Energy Supply." Economist - Innovation Summit and Awards. San Francisco: The Economist. Accessed September 23. http://www.economist.com/node/1324709.

—. 2011. Reinventing Fire: Bold Business Solutions for the New Energy Era. White River: Chelsea Green.

Lyng, Jeff, interview by John Whitcomb. 2013. Senior Policy Advisor, Center for the New Energy Economy, Colorado State University (April 24).

McDonough, William, and Michael Braungart. 2002. Cradle to Cradle: Remaking the Way We Make Things. New York City: North Point Press.

Meneveau, Charles, and Johan Meyers. 2011. "Optimal Turbine Spacing In Fully Developed Wind Farm Boundary Layers." Wind Energy: Journal (John Wiley & Sons) 15 (2): pages 305–317.

Mithun. 2009. Mithun. July 22. Accessed August 21, 2012. http://mithun.com/press/release/net-zero_energy_affordable_neighborhood_opens_on_lopez_island/.

Montreux Energy. 2012. "Clean Energy Roundtable." Aspen Roundtable. Aspen: CleanEnergy.com.

Muller, Joann. 2013. Toyota Unveils Plans for 15 New or Improved Hybrids (It Already Has 23). August 29. Accessed August 29, 2013. http://www.forbes.com/sites/joannmuller/2013/08/29/toyota-flexes-its-muscles-with-plans-for-new-wave-of-hybrids/.

National Council of Non-profits. 1997. National Council of Non-profits. Accessed June 24, 2013. www.councilofnonprofits.org.

National Grid. 2013. Naional Grid - Smart Grid. May 27. Accessed May 27, 2013. http://www.nationalgridus.com/energy/.

National Telecommunications and Information Administration. 1995. Falling Through the Net: a survey of the "have nots" in rural and urban America. July. Accessed May 31, 2013. http://www.ntia.doc.gov/ntiahome/fallingthru.html.

NaturalGas.org. 1998. Natural Gas and the Environment. Accessed May 5, 2013. http://www.naturalgas.org/environment/naturalgas.asp.

Navigant Consulting. 2013. Microgrid Deployment Tracker - 2Q. Industry Analysis, Chicago: Navigant. www/navigant.com.

Navigant Research Consulting - Asmus. 2013. Microgrid Deployment Tracker 2Q13 Commercial/Industrial, Community/Utility, Institutional/Campus, Military, and Remote Microgrids: Operating, Planned, and Proposed Projects. Q2. Accessed June 6, 2013. http://www.navigantresearch.com/research/microgrid-deployment-tracker-2q13.

Obama, Barack. 2012. State of the Union Address. January 25. Accessed January 25, 2012. http://www.whitehouse.gov/state-of-the-union-2012.

—. 2013. State of the Union Address. February 12. Accessed February 12, 2013. http://www.whitehouse.gov/the-press-office/2013/02/12/remarks-president-state-union-address.

Olgyay, Victor, and Aladar Olgyay. 1963. Design with Climate: Bioclimatic Approach to Architectural Regionalism. Princeton: Princeton University Press.

Owen, Brandon of General Electric's Global Energy Strategy & Analysis Division. 2012. "Major Clean Energy Infrastructure Investments." Clean Energy Roundtable. Aspen: Montreux Energy.

Paoletti, Dennis, interview by John Whitcomb. 2013. Acoustician (May 5).

Parker, Larry, and Mark Holt. 2007. Nuclear power: Outlook for New U.S. Reactors. Report to US Congress, Washington: Congressional Research Service (CRS).

Penny, Terry, interview by John Whitcomb. 2012. Principal Laboratory Program Manager for Advanced Vehicle and Fuel Technologies, National Renewable Energy Laboratory (June 5).

Pollock, Josh, interview by John Whitcomb. 2013. Goddard College graduate student and social media expert (April 18).

Ramirez, Frank, interview by John Whitcomb. 2013. Energy executive and entrepreneur (February 11).

Register, Richard. 2012. Ecocities– Building Cities in Balance with Nature. Vol. Fourth printing. Berkeley: New Society Publishers.

Ryan, Mike. 2012. President, PanTerra Energy. Denver, March 2.

Sears, Jim. 2012. algae@work. A2BE Carbon Capture L.L.C. May 23. Accessed May 23, 2012. http://www.algaeatwork.com/.

Senge, Peter. 2009. The Fifth Discipline - The Art and Practice of the Learning Organization. New York City: Doubleday.

Shepard, Scott. 2013. Forbes / Navigant / Pike Research - As GM Cuts Volt Price, EV Bargains Multiply. August 06. Accessed August 28, 2013. http://www.forbes.com/sites/pikeresearch/2013/08/06/as-gm-cuts-volt-price-ev-bargains-multiply/.

Sherwin, Elton. 2010. Addicted to Oil - A Venture Capitalist Perspective on How to Save our Economy and our Planet. Lexington: Energy House Publishing.

Singh, Amanjeet, Matt Syal, Sue Grady, and Sinem Korkmaz. 2010. "Effects of Green Buildings on Employee Health and Productivity." American Public Health Association.

Slaper, Timothy F. 2011. "The Triple Bottom Line: What Is It and How Does It Work?" Indiana Business Review (Indiana University).

Spensley, James "Skip", interview by John Whitcomb. 2013. University of Denver, Sturm College of Law, Adjunct Professor, State and Local Government, environmental lawyer, operations research engineer and educator (April 12).

Starbird, Michael. 2006. Meaning from Data: Statistics Made Clear (The Great Courses, Parts 1 and 2) . Chantilly: The Teaching Company.

Suroweiki, J. 2004. Wisdom of Crowds. New York City: Doubleday.

The Earth Charter Initiative. 1997. The Earth Charter. March. Accessed June 18, 2013. http://www.earthcharterinaction.org/content/pages/Read-the-Charter.html.

The Economist. 2009. Triple Bottom Line. November 17. Accessed August 20, 2013. http://www. economist.com/node/14301663.

Thompson, Paul, interview by John Whitcomb. 2013. Patent Agent, Cochran, Freund and Young L.L.C. (June 4).

Tinianow, Jerome. 2013. Chief Sustainability Officer Denver, (March 21).

Treplitz, Fran, interview by John Whitcomb. 2013. Energy Program Director, Green America (April 11).

UN-Habitat. 1997. UN-Habitat: our work. Accessed June 7, 2011. http://www.unhabitat.org/categories.asp?catid=316.

United Nations Framework Convention on Climate Change. 1998. Kyoto Protocol to the United Nations Framework - The international response to climate change. Accessed June 18, 2011. http://unfccc.int/resource/docs/convkp/kpeng.pdf.

United States Green Building Council. 2012. LEED for Neighborhood Development. Accessed January 2009. http://www.usgbc.org/resources/leed-neighborhood-development-v2009-current-version.

US Department of Energy. 2012. Sunshot Vision Study. Scientific Report, Washington: Department of Energy.

US Department of Energy, Office of Building Technology - State and Community Programs. 2008. Energy Consumption Characteristics of Commercial Building HVAC Systems. April 10. Accessed April 9, 2008. http://apps1.eere.energy.gov/buildings/publications/pdfs/commercial_initiative/hvac_volume2_final_report.pdf.

US Department of the Interior. 2012. Hydroelectric Power and Water - Basic Information About Hydroelectricity. USGS Water Sciences for Schools, Executive, United States of America, Washington D.C.: Worldwatch Institute. Accessed January.

US Energy Information Administration. 2013. Electricity Data Overview. Accessed January 7, 2013. http://www.eia.gov.

—. 2013. Energy Explained. April 14. Accessed April 14, 2013. http://www.eia.gov/electricity/.
US/Canada Task Force - Power System Outage. 2004. Final Report on the August 14, 2003

Blackout in the United States and Canada: Causes and Recommendations. Multi-Government Task Force, Ottawa / Washington D.C.: North America: Task Force.

Welch, Robert, interview by John Whitcomb. 2012. Technology Officer, TowardZero.org (September 6).

Whitcomb, John. 2012. "Bharti-Airtel Network Experience Center." Press Release. Guguaon, India: Media Centre, October 31.

Whitcomb, John. 2012, Nov, 11. Field Report: National Renewable Energy Center - Visitor Center. Field Report, Golden: Goddard.

Whitcomb, John. 2009. "Renewable Energy Power for Data Centers." Technical Report, National Renewable Energy Laboratory.

Wikipedia. 1983. Brundtland Commission. Accessed March 27, 2011. http://en.wikipedia.org/wiki/Brundtland_Commission.

Williams, Jason. 2013. "Personal Communication." Fieldwork, Atlanta.

Wise, Alison, interview by John Whitcomb. 2013. Principal, Wise Strategies (April 12).

World Commission on Environment and Development. 1987. "Our Common Future, Report of the World Commission on Environment and Development." Published as Annex to General Assembly document A/42/427, Development and International Co-operation: Environment, United Nations, New York City. Accessed June 18, 2013. http://www.un-documents.net/wced-ocf.htm.

World Nuclear Association. 2013. USA Nuclear Power. July 31. Accessed August 28, 2013. http://http://www.world-nuclear.org/info/Country-Profiles/Countries-T-Z/USA-Nuclear-Power/#.Uh5LhpLVB8E.

Worldwatch Institute. 2012. Use and Capacity of Global Hydropower. Hydropower, Washington D.C.: Worldwatch Institute.

Yergin, Daniel. 2011. The Quest: Energy, Security and the Remaking of the Modern World. New York City: Penguin Press.

Yoneda, Yuka. 2011. Tianjin Eco City is a Futuristic Green Landscape for 350,000 Residents - Green Design Will Save the World. January 10. Accessed November 21, 2011. http://inhabitat.com/tianjin-eco-city-is-a-futuristic-green-landscape-for-350000-residents/.

# Book Development Team

John Whitcomb *Author*

William E. King *Chief Editor*

Lana L. Whitcomb *Editorial Board*

Robert Welch *Technical Advisor*

Renee Forsythe *Book Design*

Ekpanith Jom Naknakorn *Graphic Artist*

Margo Whitcomb *Editor*

Chris Meehan *Editor*

Ellie Epp *Editor*

Nancy Hutchins *Editor*

Nova Cage *Publishing Services Associate*

Printed in the United States
By Bookmasters